FOUNDATIONS OF THE NONLINEAR THEORY OF ELASTICITY

OTHER *GRAYLOCK* PUBLICATIONS

FOUNDATIONS OF THE NONLINEAR THEORY OF ELASTICITY

BY

V. V. NOVOZHILOV

GRAYLOCK PRESS
ROCHESTER, N. Y.
1953

TRANSLATED FROM THE FIRST (1948) RUSSIAN EDITION

BY

F. BAGEMIHL H. KOMM W. SEIDEL

GRAYLOCK PRESS

Rochester, N. Y.

Second Printing December, 1957

Library of Congress Catalog Card Number 53-10160

Manufactured in the United States of America

CONTENTS

CHAPTER II. THE EQUILIBRIUM OF AN ELEMENT OF VOLUME OF A BODY

CHAPTER III. STRAIN ENERGY, BOUNDARY CONDITIONS, STRESS-STRAIN LAW

CHAPTER III. (Cont.)

CHAPTER IV. FORMULATION OF ELASTIC PROBLEMS IN TERMS OF STRESSES

PREFACE

This book is based on a course of lectures given by the author in 1947 in the Mathematical-Mechanical Department of the Leningrad National University. It is devoted to the exposition of the theory of elasticity without any assumptions restricting the magnitude of elongations, displacements, or angles of rotation. It also examines, in a general formulation, the connection between stresses and strains in an isotropic elastic body.

The equations of the classical (linear) theory of elasticity can be obtained from the equations of the general theory which is to be presented, by assuming that

a) the elongations and shears are negligibly small compared to unity;

b) the squares and products of the angles of rotation are negligibly small compared to the elongations and shears;

c) the connection between the stresses and strains is expressible by Hooke's law.

The nonlinear theory of elasticity, being an essential generalization of the classical theory, permits an approach to the solution of a series of important problems which do not arise in the latter theory because of its limitations. Such problems are, in particular:

1. The stability of elastic equilibrium.
2. The deformation of bodies having initial stresses.
3. The large deflection of rods.
4. The problems of torsion and bending complicated by the presence of axial forces.
5. The bending of plates and shells under deflections of the order of magnitude of the thick-

ness.

6. The deformation of elastic bodies which do not obey Hooke's law.

Finally, it has been shown recently that the equilibrium of elastic-plastic bodies (with certain restrictions) can be examined on the basis of general principles of the theory of elasticity. It is due to this fact, that the problem of the equilibrium of an elastic-plastic medium is, to a certain extent, included in the circle of problems belonging to the nonlinear theory of elasticity.

The problems listed above are very much in the forefront, and that is why ever increasing attention is being devoted to the nonlinear theory of elasticity by the scientists in the Soviet Union and other countries (see the Bibliography).

To make the book accessible to as wide a circle of readers as possible, the author has attempted to carry out all deductions in the simplest and most intuitive manner, avoiding, in particular, tensor calculus and the complicated symbolism connected with it (or, to speak more precisely, applying only the small amount of it which is given in books on the classical theory of elasticity).

In conclusion the author wishes to thank Prof. A. I. Lurye, L. M. Kachanov, Docent at Leningrad National University, and A. I. Chekmarev, the editor of this book, for some valuable critical remarks which they made when they read the manuscript.

Leningrad, November, 1947

V. V. Novozhilov

CHAPTER I

THE GEOMETRY OF STRAIN

§ 1. Coordinates

In this chapter we shall examine the basic geometrical properties of deformations (strain).

We formulate the problem as follows: Given the positions of the points of the body in its initial state (i.e., before deformation) and in its terminal state (i.e., after deformation), determine the change in the distance between two arbitrary infinitely near points of the body caused by its transition from the first state to the second. This question is purely geometrical, and neither the causes which give rise to the deformation nor the law according to which the body resists it are of any importance in its study.

Let the positions of the points of the body in its initial state be described by their projections x, y, z on the axes of some rectangular system of Cartesian coordinates X, Y, Z. Furthermore, let the points of the body undergo displacements with components u, v, w, regarded as preassigned functions of x, y, and z, along the same axes. Then the terminal position of an arbitrary point of the body is given by the Cartesian coordinates

$$\begin{aligned} \xi &= x + u(x, y, z), \\ \eta &= y + v(x, y, z), \\ \zeta &= z + w(x, y, z). \end{aligned} \tag{1.1}$$

We shall assume the functions $u, v, w,$ as well as their partial derivatives with respect to x, y, and z, to be continuous. This restriction can be called the continuity condition of the deformation.

It follows from equations (I.1) that the terminal position of the points of the body can be described in two ways: first, by the Cartesian coordinates ξ, η, ζ, and second, by the coordinates of the initial state, x, y, z. The latter, which are Cartesian coordinates for the initial state, become curvilinear coordinates for the terminal state. Indeed, if we set $x = x_0$, $y = y_0$ in (I.1), we obtain three equations

$$\begin{aligned} \xi &= x_0 + u(x_0, y_0, z), \\ \eta &= y_0 + v(x_0, y_0, z), \\ \zeta &= z + w(x_0, y_0, z), \end{aligned} \tag{I.2}$$

which together represent the curve on which those points (M_1, M_2, . . .) of the body will be situated (after deformation) which, before deforma-

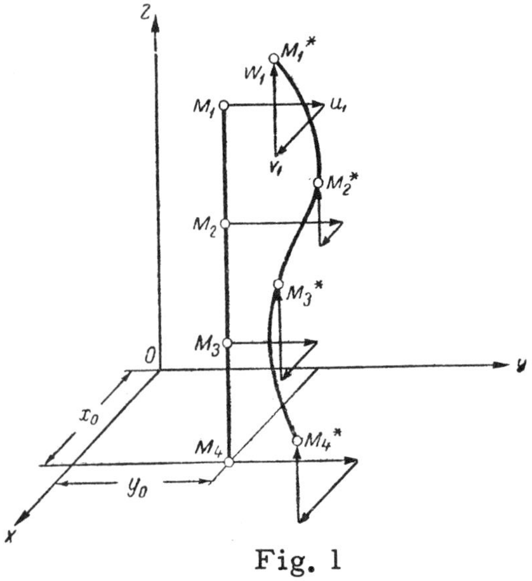

Fig. 1

tion, lay on the straight line $(x=x_0, y=y_0)$ parallel to the Z-axis (Fig. 1). Here and in the sequel, when we speak of x, y, z as curvilinear coordinates of the deformed body, we shall mark them with tildes. Clearly the coordinate lines $\tilde{x}, \tilde{y}, \tilde{z}$ in the deformed body are (generally speaking, curved) lines containing points which, before deformation, were located on straight lines parallel to the corresponding coordinate axes. This is the physical meaning of the transformation of coordinates given by equations (I.1).

The study of the deformation in the neighborhood of an arbitrary point $M(x, y, z)$ reduces to the investigation of the geometrical properties of the coordinates x, y, z regarded as the curvilinear coordinates of the deformed body.

We observe further that ξ, η, ζ, which are Cartesian coordinates for the deformed body, are curvilinear coordinates for the body before deformation. In fact, if we set $\xi=\xi_0, \eta=\eta_0$ in (I.1), we obtain three equations

$$\begin{aligned} \xi_0 &= x+u(x, y, z),\\ \eta_0 &= y+v(x, y, z),\\ \zeta &= z+w(x, y, z), \end{aligned} \tag{I.3}$$

which, when solved for x, y, z, become

$$\begin{aligned} x &= f_1(\xi_0, \eta_0, \zeta),\\ y &= f_2(\xi_0, \eta_0, \zeta),\\ z &= f_3(\xi_0, \eta_0, \zeta), \end{aligned} \tag{I.4}$$

and describe, in the system X, Y, Z, the curve on which those points of the body were situated before deformation, which, after deformation, lie on the straight line $\xi=\xi_0, \eta=\eta_0$ parallel to the Z-axis. Here and in the sequel, when ξ, η, ζ are regarded as the curvilinear coordinates of the points of the body before deformation, we shall mark them with tildes. Thus the lines $\tilde{\xi}$, $\tilde{\eta}$, $\tilde{\zeta}$ are those curves in the body before deformation, which, after deformation, become straight lines parallel to the corresponding coordinate axes. In what follows, let us agree to call the set of points of the body lying along some curve a "fiber of the body", and an infinitesimal arc of a fiber a "line element of the body". Further we shall call the set of points of the body lying on some surface a "layer of the body", and an infinitesimal element of a layer an "element of area of the body".

§2. The Angles Determining the Directions of the Coordinate Lines

As a result of the deformation, the point $M(x, y, z)$ is displaced to the position M^* having Cartesian coordinates ξ, η, ζ, whereas the point $N(x+dx, y+dy, z+dz)$ infinitely near M is displaced to the position N^* having coordinates $\xi+d\xi$, $\eta+d\eta$, $\zeta+d\zeta$.

The vector $\overrightarrow{M^*N^*}$ (with projections $d\xi$, $d\eta$, $d\zeta$) determines the magnitude and direction of that line element of the body whose magnitude and direction before deformation were given by the vector $\overrightarrow{MN}$ (with projections dx, dy, dz).

Applying formulas (I.1) to the point

$$(x+dx,\ y+dy,\ z+dz)$$

and expanding the right-hand sides in Taylor series about the point x, y, z (retaining only infinitesimals of the first order) we get

$$
\begin{aligned}
d\xi &= \left(1+\frac{\partial u}{\partial x}\right)dx+\frac{\partial u}{\partial y}dy+\frac{\partial u}{\partial z}\,dz,\\
d\eta &= \frac{\partial v}{\partial x}dx+\left(1+\frac{\partial v}{\partial y}\right)dy+\frac{\partial v}{\partial z}\,dz,\\
d\zeta &= \frac{\partial w}{\partial x}dx+\frac{\partial w}{\partial y}dy+\left(1+\frac{\partial w}{\partial z}\right)dz.
\end{aligned}
\tag{I.5}
$$

Equations (I.5) express the projections of an arbitrary line element of the body after deformation in terms of its projections before deformation.

Let us introduce the notation

$$
\begin{gathered}
e_{xx}=\frac{\partial u}{\partial x},\quad e_{yy}=\frac{\partial v}{\partial y},\quad e_{zz}=\frac{\partial w}{\partial z},\\
e_{xy}=\frac{\partial u}{\partial y}+\frac{\partial v}{\partial x},\quad e_{xz}=\frac{\partial u}{\partial z}+\frac{\partial w}{\partial x},\quad e_{yz}=\frac{\partial v}{\partial z}+\frac{\partial w}{\partial y}
\end{gathered}
\tag{I.6}
$$

and

$$
2\omega_x=\frac{\partial w}{\partial y}-\frac{\partial v}{\partial z},\quad 2\omega_y=\frac{\partial u}{\partial z}-\frac{\partial w}{\partial x},\quad 2\omega_z=\frac{\partial v}{\partial x}-\frac{\partial u}{\partial y}; \tag{I.7}
$$

then equations (I.5) can be written in the form

$$
\begin{aligned}
d\xi &= (1+e_{xx})\,dx+\left(\frac{1}{2}e_{xy}-\omega_z\right)dy+\left(\frac{1}{2}e_{xz}+\omega_y\right)dz,\\
d\eta &= \left(\frac{1}{2}e_{xy}+\omega_z\right)dx+(1+e_{yy})\,dy+\left(\frac{1}{2}e_{yz}-\omega_x\right)dz,\\
d\zeta &= \left(\frac{1}{2}e_{xz}-\omega_y\right)dx+\left(\frac{1}{2}e_{yz}+\omega_x\right)dy+(1+e_{zz})\,dz.
\end{aligned}
\tag{I.8}
$$

Let the line element before deformation be parallel to the X-axis, and have projections $(MN)_x=dx$, $(MN)_y=0$, $(MN)_z=0$. Then, according to (I.8), its projections after deformation are

$$
\left.
\begin{aligned}
d\xi=(1+e_{xx})\,dx,\quad d\eta=\left(\frac{1}{2}e_{xy}+\omega_z\right)dx,\\
d\zeta=\left(\frac{1}{2}e_{xz}-\omega_y\right)dx,
\end{aligned}
\right\}
\tag{I.9}
$$

and its length after deformation is

$$
\begin{aligned}
|M^*N^*| &= \sqrt{d\xi^2 + d\eta^2 + d\zeta^2} \\
&= \sqrt{(1+e_{xx})^2 + \left(\frac{1}{2}e_{xy} + \omega_z\right)^2 + \left(\frac{1}{2}e_{xz} - \omega_y\right)^2} \cdot dx \\
&= (1 + E_x)\, dx, \qquad (I.10)
\end{aligned}
$$

where

$$E_x = \frac{|M^*N^*| - |MN|}{|MN|} \qquad (I.11)$$

is the relative elongation of the element dx under the deformation.

Hence, the cosines of the angles formed by the vector $\overrightarrow{M^*N^*}$ with the axes X, Y, Z (in the special case where $\overrightarrow{MN}$ is parallel to the X-axis) are given by the formulas

$$
\cos(MN, X) = \frac{1+e_{xx}}{1+E_x}, \quad \cos(MN, Y) = \frac{\frac{1}{2}e_{xy} + \omega_z}{1+E_x},
$$

$$
\cos(MN, Z) = \frac{\frac{1}{2}e_{xz} - \omega_y}{1+E_x}. \qquad (I.12)
$$

As a result of the deformation, the line element dx becomes an element of arc of the line $\tilde{x}$ in the deformed body. In view of this, formulas (I.12) give the direction cosines of the tangent to this line at the point M^*. By applying analogous arguments to the line elements dy and dz, one can also derive expressions for the direction cosines of the tangents to the lines $\tilde{y}$ and $\tilde{z}$ at the same point. Let us denote the unit vectors in the directions of the above three tangents by $\mathbf{i}_1$, $\mathbf{i}_2$, $\mathbf{i}_3$ (Fig. 2). Then, in accordance with the preceding

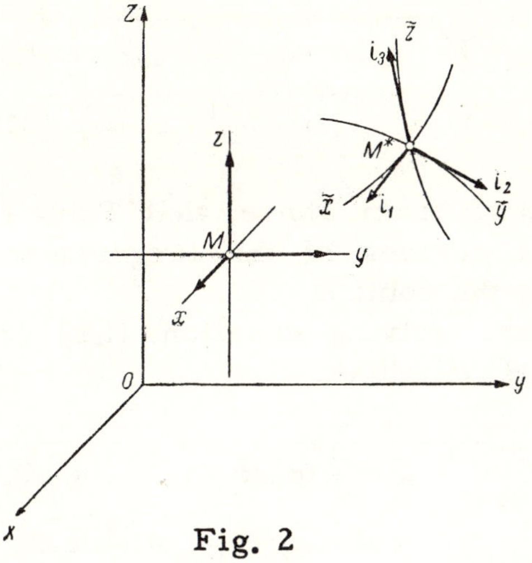

Fig. 2

results, one can construct the following table of projections of the given vectors on the axes X, Y, Z:

Table 1

	$\mathbf{i}_1$	$\mathbf{i}_2$	$\mathbf{i}_3$
X	$\frac{1+e_{xx}}{1+E_x}$	$\frac{\frac{1}{2}e_{xy}-\omega_z}{1+E_y}$	$\frac{\frac{1}{2}e_{xz}+\omega_y}{1+E_z}$
Y	$\frac{\frac{1}{2}e_{xy}+\omega_z}{1+E_x}$	$\frac{1+e_{yy}}{1+E_y}$	$\frac{\frac{1}{2}e_{yz}-\omega_x}{1+E_z}$
Z	$\frac{\frac{1}{2}e_{xz}-\omega_y}{1+E_x}$	$\frac{\frac{1}{2}e_{yz}+\omega_x}{1+E_y}$	$\frac{1+e_{zz}}{1+E_z}$

Here, in addition to the notation already introduced, we write

$$1+E_y=\sqrt{\left(\frac{1}{2}e_{xy}-\omega_z\right)^2+(1+e_{yy})^2+\left(\frac{1}{2}e_{yz}+\omega_x\right)^2},$$
$$1+E_z=\sqrt{\left(\frac{1}{2}e_{xz}+\omega_y\right)^2+\left(\frac{1}{2}e_{yz}-\omega_x\right)^2+(1+e_{zz})^2}. \quad (I.13)$$

It is clear from above, that Table 1 gives the direction cosines of the tangents to the lines $\tilde{x}$, $\tilde{y}$, $\tilde{z}$ at the point M^*.

Further, solving equations (I.8) or (I.5) for dx, dy, dz, we obtain

$$dx=\frac{1}{D}(\alpha_{11}d\xi+\alpha_{12}d\eta+\alpha_{13}d\zeta),$$
$$dy=\frac{1}{D}(\alpha_{21}d\xi+\alpha_{22}d\eta+\alpha_{23}d\zeta), \quad (I.14)$$
$$dz=\frac{1}{D}(\alpha_{31}d\xi+\alpha_{32}d\eta+\alpha_{33}d\zeta),$$

where

$$\alpha_{11}=\left(1+\frac{\partial v}{\partial y}\right)\left(1+\frac{\partial w}{\partial z}\right)-\frac{\partial v}{\partial z}\frac{\partial w}{\partial y}$$
$$=(1+e_{yy})(1+e_{zz})-\frac{1}{4}e_{yz}^2+\omega_x^2,$$
$$\alpha_{12}=\frac{\partial u}{\partial z}\frac{\partial w}{\partial y}-\left(1+\frac{\partial w}{\partial z}\right)\frac{\partial u}{\partial y}=\left(\frac{1}{2}e_{xz}+\omega_y\right)\left(\frac{1}{2}e_{yz}+\omega_x\right)$$
$$-(1+e_{zz})\left(\frac{1}{2}e_{xy}-\omega_z\right),$$
$$\alpha_{13}=\frac{\partial u}{\partial y}\frac{\partial v}{\partial z}-\left(1+\frac{\partial v}{\partial y}\right)\frac{\partial u}{\partial z}=\left(\frac{1}{2}e_{xy}-\omega_z\right)\left(\frac{1}{2}e_{yz}-\omega_x\right)$$
$$-(1+e_{yy})\left(\frac{1}{2}e_{xz}+\omega_y\right),$$
$$\alpha_{21}=\frac{\partial v}{\partial z}\frac{\partial w}{\partial x}-\left(1+\frac{\partial w}{\partial z}\right)\frac{\partial v}{\partial x}=\left(\frac{1}{2}e_{yz}-\omega_x\right)\left(\frac{1}{2}e_{xz}-\omega_y\right)$$
$$-(1+e_{zz})\left(\frac{1}{2}e_{xy}+\omega_z\right),$$
$$\alpha_{22}=\left(1+\frac{\partial u}{\partial x}\right)\left(1+\frac{\partial w}{\partial z}\right)-\frac{\partial u}{\partial z}\frac{\partial w}{\partial x}=(1+e_{xx})(1+e_{zz})$$
$$-\frac{1}{4}e_{xz}^2+\omega_y^2, \quad (I.15)$$

$$\alpha_{23}=\frac{\partial u}{\partial z}\frac{\partial v}{\partial x}-\left(1+\frac{\partial u}{\partial x}\right)\frac{\partial v}{\partial z}=\left(\frac{1}{2}e_{xz}+\omega_y\right)\left(\frac{1}{2}e_{xy}+\omega_z\right)$$
$$-(1+e_{xx})\left(\frac{1}{2}e_{yz}-\omega_x\right),$$
$$\alpha_{31}=\frac{\partial v}{\partial x}\frac{\partial w}{\partial y}-\left(1+\frac{\partial v}{\partial y}\right)\frac{\partial w}{\partial x}=\left(\frac{1}{2}e_{xy}+\omega_z\right)\left(\frac{1}{2}e_{yz}+\omega_x\right)$$
$$-(1+e_{yy})\left(\frac{1}{2}e_{xz}-\omega_y\right),$$
$$\alpha_{32}=\frac{\partial u}{\partial y}\frac{\partial w}{\partial x}-\left(1+\frac{\partial u}{\partial x}\right)\frac{\partial w}{\partial y}=\left(\frac{1}{2}e_{xy}-\omega_z\right)\left(\frac{1}{2}e_{xz}-\omega_y\right)$$
$$-(1+e_{xx})\left(\frac{1}{2}e_{yz}+\omega_x\right),$$
$$\alpha_{33}=\left(1+\frac{\partial u}{\partial x}\right)\left(1+\frac{\partial v}{\partial y}\right)-\frac{\partial u}{\partial y}\frac{\partial v}{\partial x}=(1+e_{xx})(1+e_{yy})$$
$$-\frac{1}{4}e_{xy}^2+\omega_z^2,$$

and

$$D=\begin{vmatrix}1+\frac{\partial u}{\partial x}, & \frac{\partial u}{\partial y}, & \frac{\partial u}{\partial z}\\ \frac{\partial v}{\partial x}, & 1+\frac{\partial v}{\partial y}, & \frac{\partial v}{\partial z}\\ \frac{\partial w}{\partial x}, & \frac{\partial w}{\partial y}, & 1+\frac{\partial w}{\partial z}\end{vmatrix}=\begin{vmatrix}1+e_{xx}, & \frac{1}{2}e_{xy}-\omega_z, & \frac{1}{2}e_{xz}+\omega_y\\ \frac{1}{2}e_{xy}+\omega_z, & 1+e_{yy}, & \frac{1}{2}e_{yz}-\omega_x\\ \frac{1}{2}e_{xz}-\omega_y, & \frac{1}{2}e_{yz}+\omega_x, & 1+e_{zz}\end{vmatrix}. \qquad (I.16)$$

Equations (I.14) express the projections of an arbitrary line element of the body before deformation in terms of its projections after deformation. Incidentally, in the problem under consideration, the determinant D is always different from zero, because its vanishing would signify an infinite compression of the body at the corresponding point (§11).

Let us now examine the line element $d\xi$, i. e., the line element parallel to the X-axis after deformation. According to (I.14), its projections before deformation are

$$dx=\frac{\alpha_{11}}{D}d\xi,\quad dy=\frac{\alpha_{21}}{D}d\xi,\quad dz=\frac{\alpha_{31}}{D}d\xi, \qquad (I.17)$$

and its length before deformation is

$$|MN| = \sqrt{dx^2 + dy^2 + dz^2} = \frac{1}{D}\sqrt{\alpha_{11}^2 + \alpha_{21}^2 + \alpha_{31}^2}\, d\xi = \frac{d\xi}{1 + E_\xi}, \quad (1.18)$$

where

$$E_\xi = \frac{|M^*N^*| - |MN|}{|MN|} = \frac{D}{\sqrt{\alpha_{11}^2 + \alpha_{21}^2 + \alpha_{31}^2}} - 1 \quad (1.19)$$

is the relative elongation of the element MN, which is parallel to the X-axis after deformation. Hence, in the present case, the cosines of the angles formed by the vector $\overrightarrow{MN}$ are equal to

$$\cos(MN, X) = \frac{1 + E_\xi}{D}\alpha_{11}, \quad \cos(MN, Y) = \frac{1 + E_\xi}{D}\alpha_{21},$$
$$\cos(MN, Z) = \frac{1 + E_\xi}{D}\alpha_{31}.$$

These cosines determine the direction (before deformation) of that line element passing through the point M, which, as a result of the deformation, becomes parallel to the X-axis; or, in other words, they determine the direction of the tangent to the line $\tilde{\xi}$ at the point M (see the preceding section).

By applying analogous arguments to the line elements $d\eta$ and $d\zeta$, one can also obtain the cosines of the angles which determine the directions of the tangents to the lines $\tilde{\eta}$ and $\tilde{\zeta}$ at the point M.

Let us denote the unit vectors in the directions of the above three tangents by $\mathbf{i}_1'$, $\mathbf{i}_2'$, $\mathbf{i}_3'$ (Fig. 3).

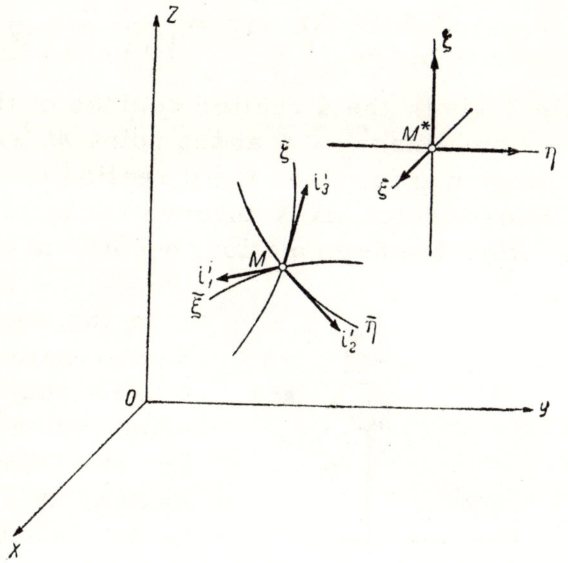

Fig. 3

Then, on the basis of the foregoing considerations, one can construct a table of the projections of these unit vectors:

Table 2

	i_1'	i_2'	i_3'
X	$\frac{1+E_\xi}{D}\alpha_{11}$	$\frac{1+E_\eta}{D}\alpha_{12}$	$\frac{1+E_\zeta}{D}\alpha_{13}$
Y	$\frac{1+E_\xi}{D}\alpha_{21}$	$\frac{1+E_\eta}{D}\alpha_{22}$	$\frac{1+E_\zeta}{D}\alpha_{23}$
Z	$\frac{1+E_\xi}{D}\alpha_{31}$	$\frac{1+E_\eta}{D}\alpha_{32}$	$\frac{1+E_\zeta}{D}\alpha_{33}$

In the table we have made use of the notation

$$E_\eta = \frac{D}{\sqrt{\alpha_{12}^2 + \alpha_{22}^2 + \alpha_{32}^2}} - 1, \quad E_\zeta = \frac{D}{\sqrt{\alpha_{13}^2 + \alpha_{23}^2 + \alpha_{33}^2}} - 1.$$

Table 2 gives the direction cosines of the tangents to the lines $\tilde{\xi}, \tilde{\eta}, \tilde{\zeta}$ at the point $M(x, y, z)$, or, in other words, gives the direction cosines of those fibers which pass through the point M and which, after the deformation, become parallel to the X-, Y-, Z-axes.

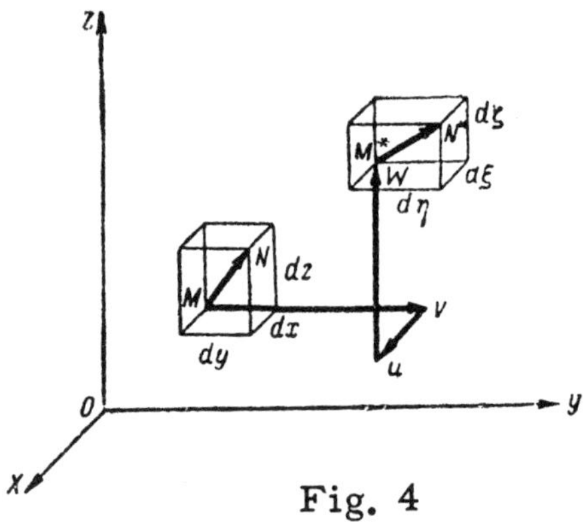

Fig. 4

In the sequel we shall examine the tensors and vectors which arise, either in the Cartesian system X, Y, Z, or in the nonorthogonal curvilinear system $\tilde{x}, \tilde{y}, \tilde{z}$, whose local unit tangent vectors are $\mathbf{i}_1, \mathbf{i}_2, \mathbf{i}_3$. Under these conditions we shall agree to denote the components in the first case by the letters ξ, η, ζ, and in the second case, by x, y, z (without tildes, to avoid complicating the formulas).

§ 3. Strain Components

The square of the distance between the points M and N (Fig. 4) before deformation is

$$ds^2 = dx^2 + dy^2 + dz^2, \tag{I.20}$$

and after deformation is

$$ds_*^2 = d\xi^2 + d\eta^2 + d\zeta^2.$$

Let us form the difference of these squares, and, in the second equation in (I.20), replace $d\xi$, $d\eta$, $d\zeta$ by their values given by formulas (I.5) or (I.8), obtaining

$$ds_*^2 - ds^2 = 2(\varepsilon_{xx}dx^2 + \varepsilon_{yy}dy^2 + \varepsilon_{zz}dz^2 + \\ + \varepsilon_{xy}dx\,dy + \varepsilon_{xz}dx\,dz + \varepsilon_{yz}dydz). \tag{I.21}$$

Here

$$\begin{aligned}
\varepsilon_{xx} &= \frac{\partial u}{\partial x} + \frac{1}{2}\left[\left(\frac{\partial u}{\partial x}\right)^2 + \left(\frac{\partial v}{\partial x}\right)^2 + \left(\frac{\partial w}{\partial x}\right)^2\right] \\
&= e_{xx} + \frac{1}{2}\left[e_{xx}^2 + \left(\frac{1}{2}e_{xy} + \omega_z\right)^2 + \left(\frac{1}{2}e_{xz} - \omega_y\right)^2\right]; \\
\varepsilon_{yy} &= \frac{\partial v}{\partial y} + \frac{1}{2}\left[\left(\frac{\partial u}{\partial y}\right)^2 + \left(\frac{\partial v}{\partial y}\right)^2 + \left(\frac{\partial w}{\partial y}\right)^2\right] \\
&= e_{yy} + \frac{1}{2}\left[e_{yy}^2 + \left(\frac{1}{2}e_{xy} - \omega_z\right)^2 + \left(\frac{1}{2}e_{yz} + \omega_x\right)^2\right]; \\
\varepsilon_{zz} &= \frac{\partial w}{\partial z} + \frac{1}{2}\left[\left(\frac{\partial u}{\partial z}\right)^2 + \left(\frac{\partial v}{\partial z}\right)^2 + \left(\frac{\partial w}{\partial z}\right)^2\right] \\
&= e_{zz} + \frac{1}{2}\left[e_{zz}^2 + \left(\frac{1}{2}e_{xz} + \omega_y\right)^2 + \left(\frac{1}{2}e_{yz} - \omega_x\right)^2\right]; \\
\varepsilon_{xy} &= \frac{\partial u}{\partial y} + \frac{\partial v}{\partial x} + \frac{\partial u}{\partial x}\frac{\partial u}{\partial y} + \frac{\partial v}{\partial x}\frac{\partial v}{\partial y} + \frac{\partial w}{\partial x}\frac{\partial w}{\partial y} \\
&= e_{xy} + e_{xx}\left(\frac{1}{2}e_{xy} - \omega_z\right) + e_{yy}\left(\frac{1}{2}e_{xy} + \omega_z\right) \\
&\quad + \left(\frac{1}{2}e_{xz} - \omega_y\right)\left(\frac{1}{2}e_{yz} + \omega_x\right); \qquad \text{(I.22)} \\
\varepsilon_{xz} &= \frac{\partial u}{\partial z} + \frac{\partial w}{\partial x} + \frac{\partial u}{\partial x}\frac{\partial u}{\partial z} + \frac{\partial v}{\partial x}\frac{\partial v}{\partial z} + \frac{\partial w}{\partial x}\frac{\partial w}{\partial z} \\
&= e_{xz} + e_{xx}\left(\frac{1}{2}e_{xz} + \omega_y\right) + e_{zz}\left(\frac{1}{2}e_{xz} - \omega_y\right) \\
&\quad + \left(\frac{1}{2}e_{xy} + \omega_z\right)\left(\frac{1}{2}e_{yz} - \omega_x\right); \\
\varepsilon_{yz} &= \frac{\partial v}{\partial z} + \frac{\partial w}{\partial y} + \frac{\partial u}{\partial y}\frac{\partial u}{\partial z} + \frac{\partial v}{\partial y}\frac{\partial v}{\partial z} + \frac{\partial w}{\partial y}\frac{\partial w}{\partial z} \\
&= e_{yz} + e_{yy}\left(\frac{1}{2}e_{yz} - \omega_x\right) + e_{zz}\left(\frac{1}{2}e_{yz} + \omega_x\right) \\
&\quad + \left(\frac{1}{2}e_{xy} - \omega_z\right)\left(\frac{1}{2}e_{xz} + \omega_y\right).
\end{aligned}$$

Introduce the notation

$$E_{MN}=\frac{ds_{*}-ds}{ds}. \tag{I.23}$$

E_{MN} is the change (produced by the deformation) in the distance between the points M and N divided by the distance between them before deformation. We call this quantity the relative elongation at the point M in the direction of the point N.

Using the notation (I.23), equation (I.21) becomes

$$E_{MN}\left(1+\frac{1}{2}E_{MN}\right)ds^2=\varepsilon_{xx}dx^2+\varepsilon_{yy}dy^2+\varepsilon_{zz}dz^2$$
$$+\varepsilon_{xy}dx\,dy+\varepsilon_{xz}dx\,dz+\varepsilon_{yz}dy\,dz \tag{I.24}$$

or

$$E_{MN}\left(1+\frac{1}{2}E_{MN}\right)=\varepsilon_{xx}\lambda^2+\varepsilon_{yy}\mu^2+\varepsilon_{zz}\nu^2$$
$$+\varepsilon_{xy}\lambda\mu+\varepsilon_{xz}\lambda\nu+\varepsilon_{yz}\mu\nu, \tag{I.25}$$

where λ, μ, and ν are the direction cosines of the vector $\overrightarrow{MN}$ (i. e., the direction of the line element before deformation)

$$\lambda=\frac{dx}{ds},\quad \mu=\frac{dy}{ds},\quad \nu=\frac{dz}{ds}. \tag{I.26}$$

We see from formula (I.25) that, in order to calculate the relative elongation at an arbitrary point M in any direction (λ, μ, ν), it is sufficient to know the six quantities ε_{xx}, ε_{yy}, ε_{zz}, ε_{xy}, ε_{xz}, ε_{yz}, which are expressible in terms of the displacements by means of equations (I.22). If the given six quantities are known at every point of the body, its deformation is completely characterized. Their vanishing at all points of the body implies the invariance of the distance between any two of its points, and this is equivalent to absence of strain. Hence it is clear that displacements u, v, w satisfying the equations

$$\varepsilon_{xx} = \varepsilon_{yy} = \varepsilon_{zz} = \varepsilon_{xy} = \varepsilon_{xz} = \varepsilon_{yz} = 0, \tag{I.27}$$

can only be displacements of a perfectly rigid body.

Let us clarify the physical meaning of the parameters $\varepsilon_{xx}, \varepsilon_{yy}, \varepsilon_{zz}, \varepsilon_{xy}, \varepsilon_{xz}, \varepsilon_{yz}$, which will be called the strain components. Setting $\mu = 0$, $\nu = 0$, $\lambda = 1$ in formula (I.25), i. e., supposing that the element under consideration was parallel to the X-axis before deformation, we obtain

$$E_x\left(1 + \frac{1}{2}E_x\right) = \varepsilon_{xx}. \tag{I.28}$$

Consequently

$$E_x = \sqrt{1 + 2\varepsilon_{xx}} - 1, \tag{I.29}$$

where E_x is the relative elongation at the point M in the direction of the X-axis. Analogously,

$$\begin{aligned} E_y &= \sqrt{1 + 2\varepsilon_{yy}} - 1, \\ E_z &= \sqrt{1 + 2\varepsilon_{zz}} - 1, \end{aligned} \tag{I.29}$$

where E_y and E_z are the relative elongations at the point M in the directions of the Y- and Z-axes, respectively.

Thus, the strain components $\varepsilon_{xx}, \varepsilon_{yy}, \varepsilon_{zz}$ characterize the elongations of those line elements which, before deformation, are parallel to the coordinate axes.

In order to clarify the physical meaning of the strain components $\varepsilon_{xy}, \varepsilon_{xz}, \varepsilon_{yz}$, let us determine the cosines of the angles which the vectors $\mathbf{i}_1, \mathbf{i}_2, \mathbf{i}_3$ form with one another (i. e., the cosines of the angles between the tangents to the lines $\tilde{x}, \tilde{y}, \tilde{z}$ passing through the point M^*).

By a well-known formula of analytic geometry,

$$\cos(\mathbf{i}_1, \mathbf{i}_2) = \cos(\mathbf{i}_1, X)\cos(\mathbf{i}_2, X) + \cos(\mathbf{i}_1, Y)\cos(\mathbf{i}_2, Y) + \cos(\mathbf{i}_1, Z)\cos(\mathbf{i}_2\, Z). \tag{I.30}$$

Replacing the cosines here by their values given in Table 1, we obtain

$$\cos(\mathbf{i}_1, \mathbf{i}_2) = \frac{\varepsilon_{xy}}{(1+E_x)(1+E_y)}. \qquad (I.31)$$

Before deformation, the angle between the line elements dx, dy is a right angle. Denote by φ_{xy} its increment due to the deformation. Then

$$\cos(\mathbf{i}_1, \mathbf{i}_2) = \cos\left(\frac{\pi}{2} - \varphi_{xy}\right)$$
$$= \sin\varphi_{xy} = \frac{\varepsilon_{xy}}{(1+E_x)(1+E_y)}. \qquad (I.32)$$

Further, denote by φ_{xz} and φ_{yz} the increments (resulting from the deformation) of the angles between the line elements dx, dz and dy, dz . Then we get two more formulas analogous to the preceding one:

$$\sin\varphi_{xz} = \frac{\varepsilon_{xz}}{(1+E_x)(1+E_z)},$$
$$\sin\varphi_{yz} = \frac{\varepsilon_{yz}}{(1+E_y)(1+E_z)}. \qquad (I.32)$$

The angles φ_{xy}, φ_{xz}, φ_{yz} are called shears.

It follows from the above equations, that the strain components ε_{xy}, ε_{xz}, ε_{yz} characterize the shears, and that if these three strain components vanish, then the angles between the line elements dx, dy, dz remain right angles after the deformation.

§ 4. Transformation of Strain Components Under Change of Axes

The same deformation can be considered in different Cartesian coordinate systems. In all such cases it will be characterized completely by the six strain components determined above, whose values, however, will depend on the choice of directions of the coordinate axes.

We are going to derive the formulas of transformation of these parameters under a change of coordinate systems. To this end we consider, together with our basic system X, Y, Z, another system X', Y', Z', the directions of whose axes relative to the axes of the first system are given in the accompanying table of cosines (Table 3).

Table 3

	X	Y	Z
X'	λ_1	μ_1	ν_1
Y'	λ_2	μ_2	ν_2
Z'	λ_3	μ_3	ν_3

Since both systems are rectangular, relations subsist between these cosines:

$$\begin{aligned} \lambda_1^2+\lambda_2^2+\lambda_3^2&=1, & \lambda_1\lambda_2+\mu_1\mu_2+\nu_1\nu_2&=0,\\ \mu_1^2+\mu_2^2+\mu_3^2&=1, & \lambda_1\lambda_3+\mu_1\mu_3+\nu_1\nu_3&=0,\\ \nu_1^2+\nu_2^2+\nu_3^2&=1, & \lambda_2\lambda_3+\mu_2\mu_3+\nu_2\nu_3&=0, \end{aligned} \tag{I.33}$$

or, written in another form,

$$\begin{aligned} \lambda_1^2+\mu_1^2+\nu_1^2&=1, & \lambda_1\mu_1+\lambda_2\mu_2+\lambda_3\mu_3&=0,\\ \lambda_2^2+\mu_2^2+\nu_2^2&=1, & \lambda_1\nu_1+\lambda_2\nu_2+\lambda_3\nu_3&=0,\\ \lambda_3^2+\mu_3^2+\nu_3^2&=1, & \mu_1\nu_1+\mu_2\nu_2+\mu_3\nu_3&=0. \end{aligned}$$

The projections on the axes of the first system, of a line element having the components dx', dy', dz' along the axes of the second system, are given by

$$\begin{aligned} dx&=\lambda_1dx'+\lambda_2dy'+\lambda_3dz',\\ dy&=\mu_1dx'+\mu_2dy'+\mu_3dz',\\ dz&=\nu_1dx'+\nu_2dy'+\nu_3dz'. \end{aligned} \tag{I.34}$$

Let us return to the equation (I.21). On the left-hand side we have the increment of the square of the distance between the points M and N, resulting from the deformation. The choice of these points is independent of the choice of the coordinate system, and therefore the left-hand side of equation (I.21) is also independent of it, and remains invariant under a change of axes.

In the right-hand side of equation (I.24), replace dx, dy, dz by their expressions in terms of the projections of the vector $\overrightarrow{MN}$ on the new coordinate axes, i. e., in terms of dx', dy', dz' (see (I.34)); we obtain

$$E_{MN}\left(1+\frac{1}{2}E_{MN}\right)ds^2 = \varepsilon'_{xx}(dx')^2 + \varepsilon'_{yy}(dy')^2 + \varepsilon'_{zz}(dz')^2 + \varepsilon'_{xy}\,dx'\,dy' + \varepsilon'_{xz}\,dx'\,dz' + \varepsilon'_{yz}\,dy'dz', \quad (I.35)$$

where

$$\begin{aligned}
\varepsilon'_{xx} &= \varepsilon_{xx}\lambda_1^2 + \varepsilon_{yy}\mu_1^2 + \varepsilon_{zz}\nu_1^2 + \varepsilon_{xy}\lambda_1\mu_1 + \varepsilon_{xz}\lambda_1\nu_1 + \varepsilon_{yz}\mu_1\nu_1,\\
\varepsilon'_{yy} &= \varepsilon_{xx}\lambda_2^2 + \varepsilon_{yy}\mu_2^2 + \varepsilon_{zz}\nu_2^2 + \varepsilon_{xy}\lambda_2\mu_2 + \varepsilon_{xz}\lambda_2\nu_2\\
&\quad + \varepsilon_{yz}\mu_2\nu_2, \qquad\qquad (I.36)\\
\varepsilon'_{zz} &= \varepsilon_{xx}\lambda_3^2 + \varepsilon_{yy}\mu_3^2 + \varepsilon_{zz}\nu_3^2 + \varepsilon_{xy}\lambda_3\mu_3 + \varepsilon_{xz}\lambda_3\nu_3 + \varepsilon_{yz}\mu_3\nu_3,\\
\varepsilon'_{xy} &= 2(\varepsilon_{xx}\lambda_1\lambda_2 + \varepsilon_{yy}\mu_1\mu_2 + \varepsilon_{zz}\nu_1\nu_2) + \varepsilon_{xy}(\lambda_1\mu_2 + \lambda_2\mu_1)\\
&\quad + \varepsilon_{xz}(\lambda_1\nu_2 + \lambda_2\nu_1) + \varepsilon_{yz}(\mu_1\nu_2 + \mu_2\nu_1),\\
\varepsilon'_{xz} &= 2(\varepsilon_{xx}\lambda_1\lambda_3 + \varepsilon_{yy}\mu_1\mu_3 + \varepsilon_{zz}\nu_1\nu_3) + \varepsilon_{xy}(\lambda_1\mu_3 + \lambda_3\mu_1)\\
&\quad + \varepsilon_{xz}(\lambda_1\nu_3 + \lambda_3\nu_1) + \varepsilon_{yz}(\mu_1\nu_3 + \mu_3\nu_1),\\
\varepsilon_{yz} &= 2(\varepsilon_{xx}\lambda_2\lambda_3 + \varepsilon_{yy}\mu_2\mu_3 + \varepsilon_{zz}\nu_2\nu_3) + \varepsilon_{xy}(\lambda_2\mu_3 + \lambda_3\mu_2)\\
&\quad + \varepsilon_{xz}(\lambda_2\nu_3 + \lambda_3\nu_2) + \varepsilon_{yz}(\mu_2\nu_3 + \mu_3\nu_2).
\end{aligned}$$

The form of (I.35) is similar to that of (I.24); it is evident that the coefficients $\varepsilon'_{xx}, \varepsilon'_{yy}, \ldots, \varepsilon'_{yz}$ entering into it have the same meaning relative to the system of axes X', Y', Z' as the coefficients $\varepsilon_{xx}, \varepsilon_{yy}, \ldots, \varepsilon_{yz}$ have relative to the system X, Y, Z.

Hence it is clear that equations (I.36) give the desired law of transformation of the strain components in passing from one coordinate system to another.

§ 5. Principal Axes of Strain

Let us find the direction for which the relative elongation E_{MN} assumes an extremal value.

If we take the X'-axis parallel to this direction, then, according to (I.35) we have

$$E_{x'}\left(1+\frac{1}{2}E_{x'}\right)=\varepsilon'_{xx} \qquad (I.37)$$

or

$$E_{x'}=\sqrt{1+2\varepsilon'_{xx}}-1.$$

We see from this expression, that the problem of determining an extremum of the elongation $E_{x'}$ reduces to the determination of an extremum of the strain component ε'_{xx}, i. e., to the determination of values λ_1, μ_1, ν_1 for which the first of the expressions (I.36) becomes extremal. Here it is a question of a relative extremum, since λ_1, μ_1, ν_1 are connected by the equation

$$\lambda_1^2+\mu_1^2+\nu_1^2-1=0. \qquad (I.38)$$

If we form the function

$$L=\varepsilon'_{xx}-\varepsilon(\lambda_1^2+\mu_1^2+\nu_1^2-1), \qquad (I.39)$$

where ε is the Lagrangian constant multiplier, and set its partial derivatives with respect to λ_1, μ_1 , and ν_1 equal to zero, we get the following three homogeneous linear equations:

$$\begin{aligned} 2(\varepsilon_{xx}-\varepsilon)\lambda_1+\varepsilon_{xy}\mu_1+\varepsilon_{xz}\nu_1&=0,\\ \varepsilon_{xy}\lambda_1+2(\varepsilon_{yy}-\varepsilon)\mu_1+\varepsilon_{yz}\nu_1&=0,\\ \varepsilon_{xz}\lambda_1+\varepsilon_{yz}\mu_1+2(\varepsilon_{zz}-\varepsilon)\nu_1&=0. \end{aligned} \qquad (I.40)$$

The trivial solution of this system is of no use

to us, because of condition (I.38), and consequently the determinant of its coefficients must equal zero:

$$\begin{vmatrix} 2(\varepsilon_{xx}-\varepsilon), & \varepsilon_{xy}, & \varepsilon_{xz} \\ \varepsilon_{xy}, & 2(\varepsilon_{yy}-\varepsilon), & \varepsilon_{yz} \\ \varepsilon_{xz}, & \varepsilon_{yz}, & 2(\varepsilon_{zz}-\varepsilon) \end{vmatrix} = 0. \qquad (I.41)$$

This equation is a cubic in the Lagrangian multiplier ε, and therefore has at least one real root, which we denote by ε_1. Note that the formula for ε'_{xx} can be rewritten as follows:

$$\varepsilon'_{xx} = (\varepsilon_{xx}\lambda_1 + \frac{1}{2}\varepsilon_{xy}\mu_1 + \frac{1}{2}\varepsilon_{xz}\nu_1)\lambda_1 + \left(\frac{1}{2}\varepsilon_{xy}\lambda_1 + \varepsilon_{yy}\mu_1 + \frac{1}{2}\varepsilon_{yz}\nu_1\right)\mu_1 + \left(\frac{1}{2}\varepsilon_{xz}\lambda_1 + \frac{1}{2}\varepsilon_{yz}\mu_1 + \varepsilon_{zz}\nu_1\right)\nu_1. \qquad (I.42)$$

If the expressions in parentheses are replaced by their values in accordance with (I.40), and we take into account (I.38), we find that the parameter ε_1 is equal to the extremal value of ε'_{xx}:

$$\varepsilon_1 = \text{Extr. } \varepsilon'_{xx}. \qquad (I.43)$$

Note, further, that the strain components ε'_{xy}, ε'_{xz} can be represented in the form

$$\varepsilon'_{xy} = 2\left[\left(\varepsilon_{xx}\lambda_1 + \frac{1}{2}\varepsilon_{xy}\mu_1 + \frac{1}{2}\varepsilon_{xz}\nu_1\right)\lambda_2 + \left(\frac{1}{2}\varepsilon_{xy}\lambda_1 + \varepsilon_{yy}\mu_1 + \frac{1}{2}\varepsilon_{yz}\nu_1\right)\mu_2 + \left(\frac{1}{2}\varepsilon_{xz}\lambda_1 + \frac{1}{2}\varepsilon_{yz}\mu_1 + \varepsilon_{zz}\nu_1\right)\nu_2\right]; \qquad (I.44)$$

$$\varepsilon'_{xz} = 2\left[\left(\varepsilon_{xx}\lambda_1 + \frac{1}{2}\varepsilon_{xy}\mu_1 + \frac{1}{2}\varepsilon_{xz}\nu_1\right)\lambda_3 + \left(\frac{1}{2}\varepsilon_{xy}\lambda_1 + \varepsilon_{yy}\mu_1 + \frac{1}{2}\varepsilon_{yz}\nu_1\right)\mu_3 + \left(\frac{1}{2}\varepsilon_{xz}\lambda_1 + \frac{1}{2}\varepsilon_{yz}\mu_1 + \varepsilon_{zz}\nu_1\right)\nu_3\right].$$

Replacing the expressions in parentheses by their values in accordance with (I.40), we obtain

$$\begin{aligned} \varepsilon'_{xy} &= 2\varepsilon_1(\lambda_1\lambda_2 + \mu_1\mu_2 + \nu_1\nu_2), \\ \varepsilon'_{xz} &= 2\varepsilon_1(\lambda_1\lambda_3 + \mu_1\mu_3 + \nu_1\nu_3), \end{aligned} \qquad (I.45)$$

but then, in view of (I.33), $\varepsilon'_{xy}=\varepsilon'_{xz}=0$. Thus, if the elongation along the X'-axis is extremal, the strain components ε'_{xy}, ε'_{xz} are equal to zero, i. e., the deformation takes place without a change in the right angles between the directions X', Y' and X', Z'.

Let us now consider another system of axes X'', Y'', Z'', which differs from X', Y', Z' by a rotation about the X'-axis through an angle θ. The table of the cosines of the angles between the axes of these two systems, Table 4, is clear from Fig. 5.

In view of the transformation formulas (I.36) and the foregoing consequences of the assumption

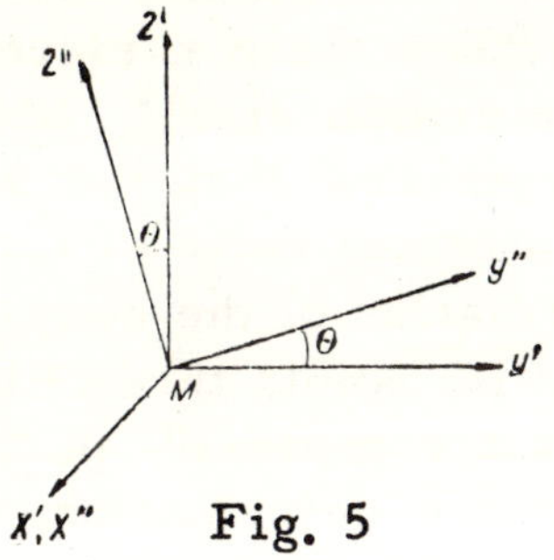

Fig. 5

Table 4

	X'	Y'	Z'
X''	1	0	0
Y''	0	$\cos\theta$	$\sin\theta$
Z''	0	$-\sin\theta$	$\cos\theta$

that the X'-axis is an extremal direction for the strain component ε'_{xx}, we have

$$\begin{aligned}
&\varepsilon''_{xx}=\varepsilon'_{xx}=\varepsilon_1,\\
&\varepsilon''_{yy}=\varepsilon'_{yy}\cos^2\theta+\varepsilon'_{zz}\sin^2\theta+\varepsilon'_{yz}\cos\theta\sin\theta,\\
&\varepsilon''_{zz}=\varepsilon'_{yy}\sin^2\theta+\varepsilon'_{zz}\cos^2\theta-\varepsilon'_{yz}\cos\theta\sin\theta,\\
&\varepsilon''_{xz}=\varepsilon''_{xy}=0,\\
&\varepsilon''_{yz}=(\varepsilon'_{zz}-\varepsilon'_{yy})\sin2\theta+\varepsilon'_{yz}\cos2\theta.
\end{aligned} \tag{I.46}$$

Let us determine the angle θ for which the elongation ε''_{yy} is extremal. A condition is obtained by differentiating ε''_{yy} with respect to θ and setting the derivative equal to zero. This yields

$$\operatorname{tg} 2\theta = \frac{\varepsilon'_{yz}}{\varepsilon'_{yy} - \varepsilon'_{zz}}. \tag{I.47}$$

It follows that whatever values ε'_{yz}, ε'_{yy}, ε'_{zz} have, one can always find an angle θ for which ε''_{yy} is extremal. For this angle, ε''_{zz} is also extremal; this is easily established by repeating the above operations for ε''_{zz}. If we substitute (I.47) in the last of equations (I.46), we see that in choosing directions Y'', Z'' from the extremality of ε''_{yy}, ε''_{zz}, the strain component ε''_{yz} vanishes.

The foregoing shows that for every point of the body one can choose three mutually perpendicular directions X'', Y'', Z'' for which the strain components ε''_{xx}, ε''_{yy}, ε''_{zz} (and consequently also the relative elongations $E_{x''}$, $E_{y''}$ $E_{z''}$) have extremal values, whereas the strain components ε''_{xy}, ε''_{xz}, ε''_{yz} (and consequently also the shears φ''_{xy}, φ''_{xz}, φ''_{yz}) are equal to zero.

Hence, given any deformation of the body, one can assign at every one of its points three fibers passing through it which are mutually perpendicular before and after the deformation. We call these three directions the principal axes of strain at the point $M(x, y, z)$, and denote the corresponding extremal values of the strain components ε_{xx}, ε_{yy}, ε_{zz} by ε_1, ε_2, ε_3.

Thus, starting from the existence of at least one extremal direction, which is a consequence of the existence of at least one real root of the cubic equation (I.41), we have proved the existence of three such directions. This shows that all three roots of equation (I.41) are always real. We shall denote by ε_1, ε_2, ε_3 the unit vectors corresponding to the principal axes of strain.

The elongations of the fibers having these directions before deformation are

$$E_1=\sqrt{1+2\varepsilon_1}-1,$$
$$E_2=\sqrt{1+2\varepsilon_2}-1, \qquad \text{(I.48)}$$
$$E_3=\sqrt{1+2\varepsilon_3}-1.$$

The cosines $(\lambda_1, \mu_1, \nu_1)$ of the angles formed by the vector ε_1 with the X-, Y-, Z-axes can be determined from the equations

$$(\varepsilon_{xx}-\varepsilon_1)\lambda_1+\frac{1}{2}\varepsilon_{xy}\mu_1+\frac{1}{2}\varepsilon_{xz}\nu_1=0,$$
$$\frac{1}{2}\varepsilon_{xy}\lambda_1+(\varepsilon_{yy}-\varepsilon_1)\mu_1+\frac{1}{2}\varepsilon_{yz}\nu_1=0, \qquad \text{(I.49)}$$
$$\frac{1}{2}\varepsilon_{xz}\lambda_1+\frac{1}{2}\varepsilon_{yz}\mu_1+(\varepsilon_{zz}-\varepsilon_1)\nu_1=0$$

and the relation

$$\lambda_1^2+\mu_1^2+\nu_1^2=1.$$

The cosines $(\lambda_2, \mu_2, \nu_2)$ and $(\lambda_3, \mu_3, \nu_3)$ of the vectors $\varepsilon_2, \varepsilon_3$ can be determined analogously.

As a result of the deformation, the fibers along the directions $\varepsilon_1, \varepsilon_2, \varepsilon_3$ which remain mutually perpendicular may undergo a certain rotation. One must therefore distinguish the principal axes of strain for the initial and terminal positions of the body.

We denote by $\varepsilon_1^*, \varepsilon_2^*, \varepsilon_3^*$ the unit vectors of the principal axes after the deformation (i. e., the directions possessed after the deformation by fibers which, before the deformation, had the directions $\varepsilon_1, \varepsilon_2, \varepsilon_3$).

The angles between the mutually perpendicular vectors $\varepsilon_1, \varepsilon_2, \varepsilon_3$ and the mutually perpendicular vectors $\varepsilon_1^*, \varepsilon_2^*, \varepsilon_3^*$ characterize the rotation which an infinitesimal element of the body about the point M undergoes as a result of the deformation. This matter will be treated more fully later on.

§ 6. Transformation of the Parameters $e_{xx}, e_{yy}, e_{zz}, e_{xy}, e_{xz}, e_{yz}$ and $\omega_x, \omega_y, \omega_z$ Under Change of Coordinate Axes

Let us determine the law of transformation of the parameters (I.6) and (I.7) under a change from the system X, Y, Z to another Cartesian system X', Y', Z', the directions of whose axes with respect to the axes of the first system are given in Table 3 (§ 4).

The components along the new axes, of the displacement of an arbitrary point of the body, are expressed in terms of its components along the old axes by the obvious formulas

$$\begin{aligned} u' &= u\lambda_1 + v\mu_1 + w\nu_1, \\ v' &= u\lambda_2 + v\mu_2 + w\nu_2, \\ w' &= u\lambda_3 + v\mu_3 + w\nu_3. \end{aligned} \tag{I.50}$$

If we substitute the first of these in the expression

$$e'_{xx} = \frac{\partial u'}{\partial x'} = \frac{\partial u'}{\partial x} \cdot \frac{\partial x}{\partial x'} + \frac{\partial u'}{\partial y}\frac{\partial y}{\partial x'} + \frac{\partial u'}{\partial z}\frac{\partial z}{\partial x'} \tag{I.51}$$

and take into account the fact that, according to (I.34),

$$\frac{\partial x}{\partial x'} = \lambda_1, \quad \frac{\partial y}{\partial x'} = \mu_1, \quad \frac{\partial z}{\partial x'} = \nu_1, \tag{I.52}$$

we get

$$\begin{aligned} e'_{xx} = e_{xx}\lambda_1^2 + e_{yy}\mu_1^2 + e_{zz}\nu_1^2 + e_{xy}\lambda_1\mu_1 \\ + e_{xz}\lambda_1\nu_1 + e_{yz}\mu_1\nu_1. \end{aligned} \tag{I.53}$$

By performing the analogous calculations for the remaining parameters (I.6), we conclude that, under a change of Cartesian coordinates, the given parameters ($e_{xx}, e_{yy}, e_{zz}, e_{xy}, e_{xz}, e_{yz}$) transform according to the same law as the strain components.

This result enables us to prove that, at every point of the body, there exist three mutually perpendicular fibers for which the parameters e_{xx}, e_{yy}, e_{zz} assume extremal values, whereas the parameters e_{xy}, e_{xz}, e_{yz} are equal to zero. Let us not dwell on this question, however, as the arguments in the present case do not differ at all from the arguments presented in the preceding section. It must be noted merely that the three fibers with the indicated property will, generally speaking, not coincide with the principal axes of strain at the point under consideration, and consequently the angles between them will not remain right angles after the deformation.

The parameters e_{xx}, e_{yy}, e_{zz} have a simple geometrical meaning. To explain it, let us examine the line element, dx, of the body, parallel (before deformation) to the X-axis. According to the first of formulas (I.8), the projection of this element (after deformation) on the same axis is

$$d\xi = (1 + e_{xx})\,dx.$$

Hence,

$$e_{xx} = \frac{d\xi - dx}{dx}, \tag{I.54}$$

i.e., e_{xx} is the relative elongation of the projection on the X-axis, of the line element whose direction before deformation was parallel to this axis. The parameters e_{yy}, e_{zz} have analogous geometrical interpretations (for the elements dy, dz). As for e_{xy}, e_{xz}, e_{yz}, they apparently do not possess an intuitive geometrical meaning in the general case of an arbitrarily large deformation.

In conclusion, let us consider the transformation formulas for $\omega_x, \omega_y, \omega_z$ under a change of coordinate axes.

In the X', Y', Z'-system, the parameter $\omega_{x'}$ is

given by the formula

$$2\omega_x' = \frac{\partial w'}{\partial y'} - \frac{\partial v'}{\partial z'} \,. \tag{I.55}$$

If we replace w' and v' by their values in accordance with (I.50), and pass from differentiation with respect to y' and z' to differentiation with respect to x, y, z, which is possible on the basis of (I.34), we obtain

$$\omega_x' = \omega_x(\mu_2\nu_3 - \nu_2\mu_3) + \omega_y(\nu_2\lambda_3 - \lambda_2\nu_3) + \omega_z(\lambda_2\mu_3 - \mu_2\lambda_3). \tag{I.56}$$

Analogously,

$$\begin{aligned} \omega_y' &= \omega_x(\nu_1\mu_3 - \mu_1\nu_3) + \omega_y(\lambda_1\nu_3 - \lambda_3\nu_1) + \omega_z(\mu_1\lambda_3 - \mu_3\lambda_1) \\ \omega_z' &= \omega_x(\mu_1\nu_2 - \mu_2\nu_1) + \omega_y(\nu_1\lambda_2 - \lambda_1\nu_2) + \omega_z(\lambda_1\mu_2 - \mu_1\lambda_2) \end{aligned} \tag{I.57}$$

These three formulas can be simplified by taking into account the fact that the unit vectors determining the directions X', Y', Z' are connected (because of the orthogonality of the axes) by the relations

$$\mathbf{i}_x' = \mathbf{i}_y' \times \mathbf{i}_z', \quad \mathbf{i}_y' = \mathbf{i}_z' \times \mathbf{i}_x', \quad \mathbf{i}_z' = \mathbf{i}_x' \times \mathbf{i}_y', \tag{I.58}$$

where $\times$ denotes the operation of vector multiplication.

Projecting (I.58) along the X, Y, Z -directions and taking Table 3 into account, we find that

$$\begin{aligned} \lambda_1 &= \mu_2\nu_3 - \nu_2\mu_3, \quad \mu_1 = \nu_2\lambda_3 - \lambda_2\nu_3, \quad \nu_1 = \lambda_2\mu_3 - \mu_2\lambda_3, \\ \lambda_2 &= \mu_3\nu_1 - \mu_1\nu_3, \quad \mu_2 = \nu_3\lambda_1 - \nu_1\lambda_3, \quad \nu_2 = \mu_1\lambda_3 - \lambda_1\mu_3, \\ \lambda_3 &= \mu_1\nu_2 - \mu_2\nu_1, \quad \mu_3 = \nu_1\lambda_2 - \lambda_1\nu_2, \quad \nu_3 = \lambda_1\mu_2 - \mu_1\lambda_2. \end{aligned} \tag{I.59}$$

These are well-known relations between the direction cosines of two Cartesian systems of rectangular coordinates on the assumption that both systems have the same orientation of axes. If one of the systems is right-handed and the other is left-handed, it is necessary to rearrange the terms appearing in the right-hand members.

By (I.59) and the foregoing remark, formulas

(I.56, I.57) assume the form

$$\omega_x' = \pm(\omega_x\lambda_1 + \omega_y\mu_1 + \omega_z\nu_1),$$
$$\omega_y' = \pm(\omega_x\lambda_2 + \omega_y\mu_2 + \omega_z\nu_2), \qquad (I.60)$$
$$\omega_z' = \pm(\omega_x\lambda_3 + \omega_y\mu_3 + \omega_z\nu_3),$$

where the upper sign must be taken if the systems X, Y, Z and X', Y', Z' have the same orientation, the lower sign if different.

This shows that, under a change of coordinates, the parameters ω_x, ω_y, ω_z transform as the projections of the axial vector ω whose length is

$$\omega = \sqrt{\omega_x^2 + \omega_y^2 + \omega_z^2}, \qquad (I.61)$$

and whose direction is given by the cosines

$$\cos(\omega, X) = \frac{\omega_x}{\omega}, \quad \cos(\omega, Y) = \frac{\omega_y}{\omega}, \quad \cos(\omega, Z) = \frac{\omega_z}{\omega}. \qquad (I.62)$$

§ 7. Geometrical Meaning of the Parameters ω_x, ω_y, ω_z

Let us imagine the point M to coincide with the point M^*, and transfer the origin of the coordinate system X, Y, Z to this common point (without changing the directions of the axes). Then the vectors $\overrightarrow{MN}$, $\overrightarrow{M^*N^*}$ both issue from the origin of coordinates and have the projections

$$\overrightarrow{MN} \sim dx,\ dy,\ dz,$$
$$\overrightarrow{M^*N^*} \sim d\xi,\ d\eta,\ d\zeta.$$

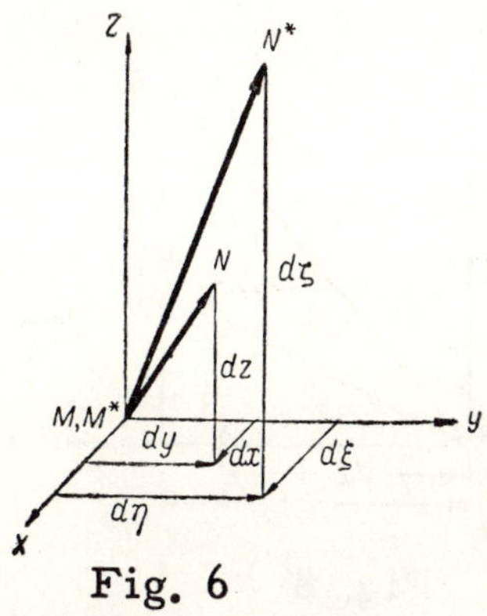

Fig. 6

Such a representation does away with the unessential translation of the neighborhood of the point M, thus enabling us to concentrate our attention on the comparison of the magnitudes and directions of the vectors

$\overrightarrow{MN}$ and $\overrightarrow{M^*N^*}$ describing the relative positions of the points M and N before and after the deformation.

The magnitudes which determine the increments of the lengths of the line elements of the body, and the increments of the angles between them, resulting from the deformation, have already been studied in the preceding sections. Under a deformation, however, not only do the relative directions of the fibers change, but also their absolute directions. In view of this, an infinitesimal element of volume of the body in its initial position undergoes a certain rotation, in addition to a deformation, in passing to the terminal position. The term rotation, as applied to an element of volume which, in the process of displacement, alters not only its position but also its dimensions and form, will be understood to represent the mean value of the rotations experienced by the totality of line elements belonging to the given element of volume.

Let us agree to regard as the angle of rotation of a fiber about an axis Ξ, to which it is perpendicular before the deformation, the angle ψ between this fiber (in its position before deformation)

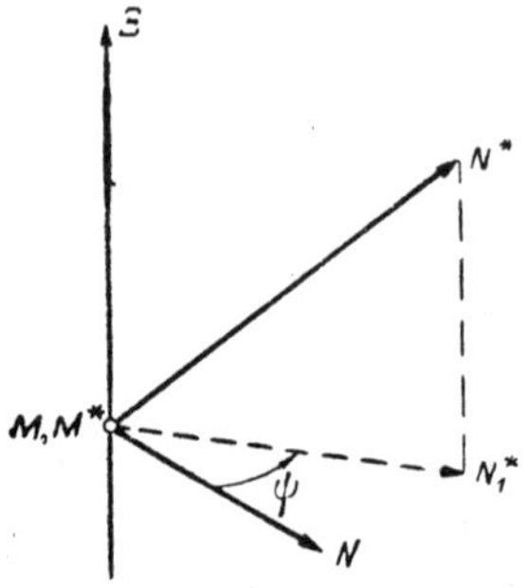

Fig. 7

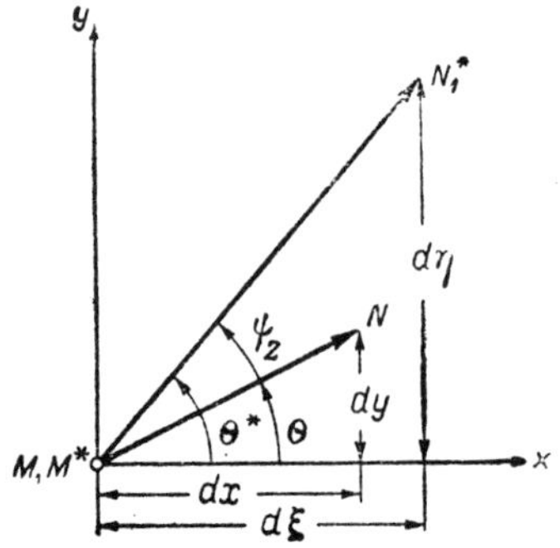

Fig. 8

and its projection (after deformation) on a plane which is perpendicular to the given axis (Fig. 7).

To clarify the magnitudes characterizing the rotation which a neighborhood of the point M undergoes as a result of the displacements u, v, w, we turn to formulas (I.8) and apply them to the special case where the line element MN is perpendicular to the Z-axis. Then $dz=0$, and equations (I.8) assume the form

$$\begin{aligned} d\xi &= (1+e_{xx})\,dx + \left(\frac{1}{2}e_{xy} - \omega_z\right)dy, \\ d\eta &= \left(\frac{1}{2}e_{xy} + \omega_z\right)dx + (1+e_{yy})\,dy, \\ d\zeta &= \left(\frac{1}{2}e_{xz} - \omega_y\right)dx + \left(\frac{1}{2}e_{yz} + \omega_x\right)dy. \end{aligned} \tag{I.63}$$

Fig. 8 shows the XY-plane, where the segment MN represents the line element before the deformation (the line element in its natural length), and the segment MN_1^* is the projection of M^*N^* on the plane under consideration.

From the figure it is clear that

$$tg\theta = \frac{dy}{dx}, \qquad tg\theta^* = \frac{d\eta}{d\xi}. \tag{I.64}$$

In the second of these equations, replace $d\xi$ and $d\eta$ by their values in (I.63), obtaining

$$tg\theta^* = \frac{\left(\frac{1}{2}e_{xy} + \omega_z\right)\cos\theta + (1+e_{yy})\sin\theta}{(1+e_{xx})\cos\theta + \left(\frac{1}{2}e_{xy} - \omega_z\right)\sin\theta}. \tag{I.65}$$

Let us examine the angle

$$\psi_z = \theta^* - \theta, \tag{I.66}$$

i. e., the angle of rotation of the fiber MN about the Z-axis, due to the deformation. According to (I.65) and (I.66),

$$\frac{\operatorname{tg}\theta+\operatorname{tg}\psi_z}{1-\operatorname{tg}\theta\operatorname{tg}\psi_z}=\frac{\left(\frac{1}{2}e_{xy}+\omega_z\right)\cos\theta+(1+e_{yy})\sin\theta}{(1+e_{xx})\cos\theta+\left(\frac{1}{2}e_{xy}-\omega_z\right)\sin\theta}. \qquad (I.67)$$

Hence,

$$\operatorname{tg}\psi_z=\frac{\omega_z+\frac{1}{2}e_{xy}\cos 2\theta+\frac{1}{2}(e_{yy}-e_{xx})\sin 2\theta}{1+e_{xx}\cos^2\theta+e_{yy}\sin^2\theta+\frac{1}{2}e_{xy}\sin 2\theta}. \qquad (I.68)$$

The mean value of $\operatorname{tg}\psi_z$ in the interval from $\theta=0$ to $\theta=2\pi$ (i. e., its mean value for all the fibers perpendicular to the Z-axis before the deformation) is given by the expression

$$\overline{\operatorname{tg}\psi_z}=\frac{1}{2\pi}\int_0^{2\pi}\operatorname{tg}\psi_z d\theta=I_1+I_2. \qquad (I.69)$$

Here

$$I_1=\frac{\omega_z}{2\pi}\int_0^{2\pi}\frac{d\theta}{1+e_{xx}\cos^2\theta+e_{yy}\sin^2\theta+\frac{1}{2}e_{xy}\sin 2\theta},$$

$$I_2=\frac{1}{4\pi}\int_0^{2\pi}\frac{e_{xy}\cos 2\theta+(e_{yy}-e_{xx})\sin 2\theta}{1+e_{xx}\cos^2\theta+e_{yy}\sin^2\theta+\frac{1}{2}e_{xy}\sin 2\theta}d\theta. \qquad (I.70)$$

The integral I_2 can be evaluated by making the substitution

$$f=1+e_{xx}\cos^2\theta+e_{yy}\sin^2\theta+\frac{1}{2}e_{xy}\sin 2\theta, \qquad (I.71)$$

which yields

$$I_2=\frac{1}{4\pi}\int_{\theta=0}^{\theta=2\pi}\frac{df}{f}=\frac{1}{4\pi}\left|\ln f\right|_{\theta=0}^{\theta=2\pi}=0. \qquad (I.72)$$

The integral I_1 is reducible to the form

$$I_1 = \frac{\omega_z}{\pi}\int_0^{2\pi} \frac{d\theta}{2+e_{xx}+e_{yy}+\sqrt{(e_{xx}-e_{yy})^2+e^2_{xy}}\sin(2\theta+\beta)}$$
$$= \frac{\omega_z}{2\pi}\int_{\beta}^{4\pi+\beta} \frac{d\gamma}{2+e_{xx}+e_{yy}+\sqrt{(e_{xx}-e_{yy})^2+e^2_{xy}}\sin\gamma}, \qquad (I.73)$$

where

$$\beta = \arcsin\left(\frac{e_{xx}-e_{yy}}{\sqrt{(e_{xx}-e_{yy})^2+e^2_{xy}}}\right)$$
$$= \arccos\left(\frac{e_{xy}}{\sqrt{(e_{xx}-e_{yy})^2+e^2_{xy}}}\right) \qquad (I.74)$$

Hence,

$$\overline{\operatorname{tg}\psi_z} = \frac{1}{2\pi}\frac{\omega_z}{\sqrt{1+e_{xx}+e_{yy}+e_{xx}e_{yy}-\frac{1}{4}e^2_{xy}}}$$
$$\times \left|\operatorname{arctg}\frac{(2+e_{xx}+e_{yy})\operatorname{tg}\frac{\gamma}{2}+\sqrt{(e_{xx}-e_{yy})^2+e^2_{xy}}}{2\sqrt{1+e_{xx}+e_{yy}+e_{xx}e_{yy}-\frac{1}{4}e^2_{xy}}}\right|_{\beta}^{4\pi+\beta} \qquad (I.75)$$

Since the function arctg ( ) is multiple-valued, the result obtained is indefinite. This indefiniteness, however, can be removed by taking into account the fact that as e_{xx}, e_{yy}, e_{xy} tend to zero, the integral I_1 (and therefore also $\overline{\operatorname{tg}\psi_z}$) must tend to ω_z, as can be seen from (I.73). Consequently in (I.75) we must set

$$\left|\operatorname{arctg}\frac{(2+e_{xx}+e_{yy})\operatorname{tg}\frac{\gamma}{2}+\sqrt{(e_{xx}-e_{yy})^2+e^2_{xy}}}{2\sqrt{1+e_{xx}+e_{yy}+e_{xx}e_{yy}-\frac{1}{4}e^2_{xy}}}\right|_{\beta}^{4\pi+\beta} = 2\pi, \qquad (I.76)$$

which leads to the following expression for $\overline{\operatorname{tg}\psi_z}$:

$$\overline{\operatorname{tg}\psi_z} = \frac{\omega_z}{\sqrt{(1+e_{xx})(1+e_{yy})-\frac{1}{4}e^2_{xy}}}. \qquad (I.77)$$

Analogously one can write down two more formulas

$$\overline{\operatorname{tg}\psi_x} = \frac{\omega_x}{\sqrt{(1+e_{yy})(1+e_{zz}) - \frac{1}{4}e_{yz}^2}},$$
$$\overline{\operatorname{tg}\psi_y} = \frac{\omega_y}{\sqrt{(1+e_{xx})(1+e_{zz}) - \frac{1}{4}e_{xz}^2}},\qquad (I.77)$$

which determine the mean values of the tangents of the angles of rotation about the X-and Y-axes, of the line elements of the body perpendicular to these axes before the deformation.

The three parameters $\overline{\operatorname{tg}\psi_x}$, $\overline{\operatorname{tg}\psi_y}$, $\overline{\operatorname{tg}\psi_z}$ characterize the rotation of an infinitesimal volume containing the point M; they are proportional to ω_x, ω_y, and ω_z, and vanish whenever these parameters are equal to zero.

It is clear from (I.60) that if ω_x, ω_y, ω_z are equal to zero in some coordinate system X, Y, Z, then they are equal to zero in any other coordinate system. It follows that if the relations

$$\omega_x = \omega_y = \omega_z = 0 \qquad (I.78)$$

hold at some point of the body, then, in the mean, the line elements passing through this point will not undergo a rotation relative to any axis passing through this point.

Thus, relations (I.78) are the conditions for the absence of rotation of an arbitrary infinitesimal element of volume of the body, containing the point M.

§ 8. Fibers Preserving Direction Under Deformation

Let us now approach the conditions for absence of rotation, in a different way, and, in the course of the argument, establish the fact, which is interesting in itself, that at every point in the body there exists at least one fiber which preserves its direction under a deformation. For such a fiber the vectors $\overrightarrow{MN}$ and $\overrightarrow{M^*N^*}$ (Fig. 6) are identical in direction, which implies that their projections satisfy the relations

$$\frac{d\xi}{dx}=\frac{d\eta}{dy}=\frac{d\zeta}{dz}=\chi=\text{const.}, \tag{I.79}$$

where

$$\chi=\frac{|M^*N^*|}{|MN|}=1+E', \tag{I.80}$$

and E' is the elongation in the direction MN.

Substituting the values of $d\xi$, $d\eta$, $d\zeta$ from (I.8) in (I.79), we obtain the following linear algebraic system:

$$\begin{aligned}
&(e_{xx}-E')\lambda+\left(\frac{1}{2}e_{xy}-\omega_z\right)\mu+\left(\frac{1}{2}e_{xz}+\omega_y\right)\nu=0,\\
&\left(\frac{1}{2}e_{xy}+\omega_z\right)\lambda+(e_{yy}-E')\mu+\left(\frac{1}{2}e_{yz}-\omega_x\right)\nu=0,\\
&\left(\frac{1}{2}e_{xz}-\omega_y\right)\lambda+\left(\frac{1}{2}e_{yz}+\omega_x\right)\mu+(e_{zz}-E')\nu=0,
\end{aligned} \tag{I.81}$$

in which

$$\lambda=\frac{dx}{|MN|},\quad \mu=\frac{dy}{|MN|},\quad \nu=\frac{dz}{|MN|} \tag{I.82}$$

are the direction cosines of the desired fiber which does not rotate under the deformation. Since these cosines satisfy the condition

$$\lambda^2+\mu^2+\nu^2=1, \tag{I.83}$$

the trivial solution of the system (I.81) is not the solution of our problem. It is therefore necessary to equate the determinant of the coefficients of the system (I.81) to zero, which leads to the equation

$$\begin{vmatrix} e_{xx}-E', & \frac{1}{2}e_{xy}-\omega_z, & \frac{1}{2}e_{xz}+\omega_y \\ \frac{1}{2}e_{xy}+\omega_z, & e_{yy}-E', & \frac{1}{2}e_{yz}-\omega_x \\ \frac{1}{2}e_{xz}-\omega_y, & \frac{1}{2}e_{yz}+\omega_x, & e_{zz}-E' \end{vmatrix} = 0. \tag{I.84}$$

It is shown in higher algebra that a sufficient condition for the three roots of an equation of type (I.84) to be real, is the symmetry of the elements of the determinant with respect to the main diagonal. This is what guarantees that the roots of equation (I.41) are real.

The determinant (I.84), unlike (I.41), is not symmetric, and so, in general, equation (I.84) will have only one real root E_1'. By substituting this root in equations (I.81) and adjoining relation (I.83) to these, we can determine λ_1, μ_1, ν_1, the direction cosines of the invariant fiber.

If certain relations subsist between the elements of the determinant, equation (I.84) may have three real roots. Without investigating this question in detail, we observe (as a rough estimate) that this can take place only if the maximal rotations do not exceed the maximal shears. If the rotation at some point of the body satisfies this condition, then there may be three fibers at this point which remain invariant in direction under the deformation, but in general they will not be mutually perpendicular. Their directions are determined from equations (I.81) and relations of the form (I.83), as was explained earlier.

In conclusion, let us consider the special case where

$$\omega_x = \omega_y = \omega_z = 0. \tag{I.85}$$

Under these conditions the determinant (I.84) is symmetric, and its three roots are real. The analogy here between equations (I.41) and (I.84), together with the identity of the laws of transformation of the parameters $e_{xx}, e_{yy}, \ldots, e_{yz}$ and of the strain components, enables us to assert that if the relations (I.85) hold at a point of the body, then there exist three mutually perpendicular fibers at this point, whose directions are invariant under the deformation. Since these fibers are also mutually perpendicular after the deformation, they must coincide with the principal axes of strain at the point in question.

The vectors $\varepsilon_1, \varepsilon_2, \varepsilon_3$, in turn, (which coincide with the directions of the invariant fibers), are equal to the vectors $\varepsilon_1^*, \varepsilon_2^*, \varepsilon_3^*$; i. e., in the present case, the principal axes in the deformed body will coincide with the principal axes in the body before deformation.

We have thus arrived again at the conclusion that relations (I.85) are the conditions for the absence of rotation in the neighborhood of a point where they subsist. It is customary to call a deformation without rotation, a pure strain. Under such a deformation, the fibers in the directions of the vectors $\varepsilon_1, \varepsilon_2, \varepsilon_3$ are elongated, but preserve their initial directions.

§ 9. Invariants of Strain and Rotation

If we expand the determinant (I.41), we get

$$\varepsilon^3 - a_2\varepsilon^2 + a_1\varepsilon - a_0 = 0, \qquad (I.86)$$

where

$$a_2 = \varepsilon_{xx} + \varepsilon_{yy} + \varepsilon_{zz},$$
$$a_1 = \varepsilon_{xx}\varepsilon_{yy} + \varepsilon_{xx}\varepsilon_{zz} + \varepsilon_{yy}\varepsilon_{zz} - \frac{1}{4}\left(\varepsilon_{xy}^2 + \varepsilon_{xz}^2 + \varepsilon_{yz}^2\right), \qquad (I.87)$$
$$a_0 = \begin{vmatrix} \varepsilon_{xx}, & \frac{1}{2}\varepsilon_{xy}, & \frac{1}{2}\varepsilon_{xz} \\ \frac{1}{2}\varepsilon_{xy}, & \varepsilon_{yy}, & \frac{1}{2}\varepsilon_{yz} \\ \frac{1}{2}\varepsilon_{xz}, & \frac{1}{2}\varepsilon_{yz}, & \varepsilon_{zz} \end{vmatrix} = \varepsilon_{xx}\varepsilon_{yy}\varepsilon_{zz} - \frac{1}{4}\left(\varepsilon_{xx}\varepsilon_{yz}^2 + \varepsilon_{yy}\varepsilon_{xz}^2 + \varepsilon_{zz}\varepsilon_{xy}^2 - \varepsilon_{xy}\varepsilon_{xz}\varepsilon_{yz}\right).$$

Since the roots (ε_1, ε_2, ε_3) of this equation are independent of the choice of directions X, Y, Z, the coefficients a_0, a_1, a_2 are independent of this choice, and so they turn out to be invariants of the deformation.

By well-known relations subsisting between the coefficients and roots of algebraic equations, the coefficients are

$$a_2 = \varepsilon_1 + \varepsilon_2 + \varepsilon_3,$$
$$a_1 = \varepsilon_1\varepsilon_2 + \varepsilon_1\varepsilon_3 + \varepsilon_2\varepsilon_3, \qquad (I.88)$$
$$a_0 = \varepsilon_1\varepsilon_2\varepsilon_3.$$

Expanding (I.84), we have, further,

$$(E')^3 - b_2(E')^2 + b_1E' - b_0 = 0, \qquad (I.89)$$

where

$$b_2 = e_{xx} + e_{yy} + e_{zz},$$
$$b_1 = e_{xx}e_{yy} + e_{xx}e_{zz} + e_{yy}e_{zz} - \frac{1}{4}\left(e_{xy}^2 + e_{xz}^2 + e_{yz}^2\right) + \omega_x^2 + \omega_y^2 + \omega_z^2, \qquad (I.90)$$
$$b_0 = \begin{vmatrix} e_{xx}, & \frac{1}{2}e_{xy} - \omega_z, & \frac{1}{2}e_{xz} + \omega_y \\ \frac{1}{2}e_{xy} + \omega_z, & e_{yy}, & \frac{1}{2}e_{yz} - \omega_x \\ \frac{1}{2}e_{xz} - \omega_y, & \frac{1}{2}e_{yz} + \omega_x, & e_{zz} \end{vmatrix} = e_{xx}e_{yy}e_{zz}$$
$$+ \frac{1}{4}\left(e_{xy}e_{yz}e_{xz} - e_{yy}e_{xz}^2 - e_{zz}e_{xy}^2 - e_{xx}e_{yz}^2\right) + \omega_x^2 e_{xx} + \omega_y^2 e_{yy} + e_{zz}\omega_z^2 + \omega_x\omega_y e_{xy} + \omega_x\omega_z e_{xz} + \omega_y\omega_z e_{yz}.$$

These quantities, as well as those in (I.87), remain invariant under a transformation of coordinates. Remembering the analogy between the laws of transformation of the strain components and of the parameters $e_{xx}, e_{yy}, \ldots, e_{yz}$, we can assert that these parameters must possess the invariants

$$
\begin{aligned}
b_2' &= b_2 = e_{xx} + e_{yy} + e_{zz}, \\
b_1' &= e_{xx} e_{yy} + e_{xx} e_{zz} + e_{yy} e_{zz} - \frac{1}{4}\left(e_{xy}^2 + e_{xz}^2 + e_{yz}^2\right), \\
b_0' &= e_{xx} e_{yy} e_{zz} + \frac{1}{4}\left(e_{xy} e_{xz} e_{yz} - e_{xx} e_{yz}^2 - e_{yy} e_{xz}^2 \right. \\
&\qquad \left. - e_{zz} e_{xy}^2\right),
\end{aligned}
\tag{I.91}
$$

so that the expressions

$$
\begin{aligned}
b_1'' &= b_1 - b_1' = \omega_x^2 + \omega_y^2 + \omega_z^2, \\
b_0'' &= b_0 - b_0' = \omega_x^2 e_{xx} + \omega_y^2 e_{yy} + \omega_z^2 e_{zz} \\
&\qquad + \omega_x \omega_y e_{xy} + \omega_x \omega_z e_{xz} + \omega_y \omega_z e_{yz}
\end{aligned}
\tag{I.92}
$$

must also be invariants.

The invariance of the first of these was already proved in § 6.

§ 10. The General Picture of the Deformation in the Neighborhood of an Arbitrary Point of the Body

It follows from (I.8) that the projections of the vector $\overrightarrow{M^*N^*}$ (i. e., the projections of an arbitrary line element of the body after deformation) are connected by means of linear relations with the projections of the vector $\overrightarrow{MN}$ (i. e., with the projections of the same element before deformation). Correspondingly, the inverse relations expressible by (I.14) are also linear. The coefficients in (I.8) and (I.14) are to be taken constant and equal to their values at the point M, since taking their variability into account would be equivalent to taking into account infinitesimals of higher order in (I.8) and (I.14). Thus the deformation of an infinitesimal region containing the point M is described by a linear transformation with constant coefficients.

As is well known from analytic geometry, however, such a transformation preserves straight lines, planes, and parallelism of straight lines and planes. Such a transformation, moreover, preserves quadric surfaces. In particular, a sphere is transformed into an ellipsoid (triaxial in general). It follows that, under the deformation, every infinitesimal parallelopiped in the body before deformation becomes a parallelopiped with different side lengths and different angles between the sides (in general).

In particular, the rectangular parallelopiped with edges dx, dy, dz parallel to the coordinate axes is transformed by the deformation into an oblique parallelopiped with edges $(1+E_x)\,dx$,

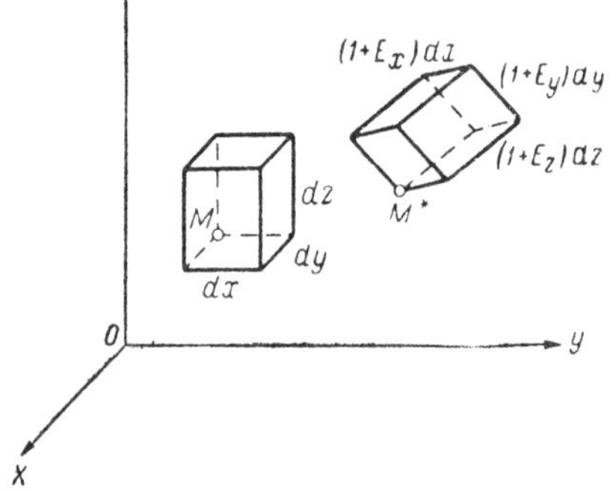

Fig. 9

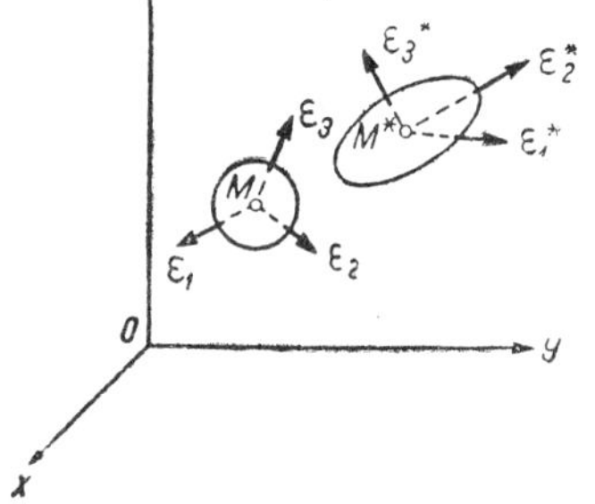

Fig. 10

$(1+E_y)\,dy$, $(1+E_z)\,dz$ forming angles $\frac{\pi}{2}-\varphi_{xy}$, $\frac{\pi}{2}-\varphi_{xz}$, $\frac{\pi}{2}-\varphi_{yz}$ (Fig. 9). The parallelopiped whose edges before the deformation coincide with the principal axes at the point in question is still rectangular after the deformation, and has edges $(1+E_1)\,da$, $(1+E_2)\,db$, $(1+E_3)\,dc$, where da, db, dc are the lengths of the edges before the deformation.

The points which lie on a sphere of radius ds before the deformation, lie on an ellipsoid with semi-axes $(1+E_1)\,ds$, $(1+E_2)\,ds$, $(1+E_3)\,ds$ after the deformation. The directions of the axes of this ellipsoid coincide with $\overset{*}{\varepsilon}_1, \overset{*}{\varepsilon}_2, \overset{*}{\varepsilon}_3$, i. e., with the principal axes in the deformed body, because the extremal values of the elongations evidently correspond to the semi-axes of the ellipsoid (Fig. 10).

The foregoing gives some idea of the character of the deformation of an infinitesimal region surrounding the point M. One can say that, under a deformation, this region first undergoes a translation, as a result of which the point M coincides with the point M^*; secondly, experiences a rotation, under which the fibers directed along ε_1, ε_2, ε_3 become directed along $\overset{*}{\varepsilon}_1, \overset{*}{\varepsilon}_2, \overset{*}{\varepsilon}_3$; and finally, undergoes a pure strain, under which the fibers $\overset{*}{\varepsilon}_1, \overset{*}{\varepsilon}_2, \overset{*}{\varepsilon}_3$ receive elongations E_1, E_2, E_3.

In the preceding sections we have presented all the information necessary for actually calculating (for given displacements u, v, w) the three factors listed above, which determine the final position and shape of an infinitesimal element of volume in the body.

We remark that the translation and rotation of an element of volume are not the characteristics of its deformation. The latter is determined solely by the components ε_{xx}, ε_{yy}, ε_{zz}, ε_{xy}, ε_{xz}, ε_{yz}.

If one speaks of the deformation of the body as a whole, however, then precisely the displacements of its points and the angles of rotation of its fibers are more characteristic.

Thus, e. g., the deformation of a beam is usually taken to mean its deflection (i. e., its displacement), whereas the deformation of a shaft is understood to be the twisting of one of its ends relative to the other (i. e., the angle of rotation).

From this standpoint, displacements and rotations can be called the characteristics of the deformation of a body as a whole, whereas elongations and shears can be called the characteristics of the deformation of an infinitesimal element of volume of the body.

The indicated duality of the notion of deformation inherent in the necessity of distinguishing its micro- and macrocharacteristics gives rise to the possibility of two different interpretations of the term "small deformation". According to the above, it can be understood to mean either the smallness of the elongations and shears (compared to unity) or the smallness of the displacements (compared to the linear dimensions of the body) and angles of rotation (compared to unity).

These two definitions must not be confused. In some books on the theory of elasticity, however, they are confused. The fact is that the classical theory of small deformations actually rests on the assumption that the displacements and rotations are small, but this is rarely made sufficiently explicit. As a result, the reader, who is accustomed to associate the notion of a deformation with its components, readily gets the impression that the smallness of the elongations and shears is meant. It should be emphasized that the assumption that the displacements and rotations are small is a greater restriction of the generality of the arguments than the assumption that the strain components are small. The first assumption implies the second, but the converse is false. This will be clarified later. It must also be remarked that, in those cases where the necessity of small displacements is indicated, it is ordinarily not specified what they must be small in comparison with. Such a specification, however, is absolutely necessary, since displacements are dimensional quantities.

To avoid this situation, some authors resort to the notion of an infinitesimal displacement, making it the foundation of their considerations. This solution of the question, however, is unsatisfactory, because it is not at all clear within what limits one can apply the theory of elasticity when it rests on such a purely formal basis.

The methodological inaccuracies just enumerated (which are characteristic of many books on the theory of elasticity) deserve to be mentioned, because they concern the foundations of our subject.

As for ourselves, when we use the term "small deformation" in the sequel, we shall always mean the smallness of the elongations and shears compared to unity. If, in addition, the rotations and displacements in the problems examined are also small, then this will always be indicated explicitly.

§ 11. Change in Volume

An infinitesimal parallelopiped with edges dx, dy, dz becomes an oblique parallelopiped after the deformation, the projections of whose edges on the X-, Y-, Z-axes, in view of (I.5), are equal to

$$\left(1+\frac{\partial u}{\partial x}\right)dx,\ \frac{\partial u}{\partial y}dy,\ \frac{\partial u}{\partial z}dz \qquad \text{on the } X\text{-axis},$$

$$\frac{\partial v}{\partial x}dx,\ \left(1+\frac{\partial v}{\partial y}\right)dy,\ \frac{\partial v}{\partial z}dz \qquad \text{on the } Y\text{-axis},$$

$$\frac{\partial w}{\partial x}dx,\ \frac{\partial w}{\partial y}dy,\ \left(1+\frac{\partial w}{\partial z}\right)dz \qquad \text{on the } Z\text{-axis}.$$

Hence, using a well-known formula of analytic geometry, we obtain

$$V^* = \begin{vmatrix} 1+\frac{\partial u}{\partial x}, & \frac{\partial u}{\partial y}, & \frac{\partial u}{\partial z} \\ \frac{\partial v}{\partial x}, & 1+\frac{\partial v}{\partial y}, & \frac{\partial v}{\partial z} \\ \frac{\partial w}{\partial x}, & \frac{\partial w}{\partial y}, & 1+\frac{\partial w}{\partial z} \end{vmatrix} dxdydz = Ddxdydz, \tag{I.93}$$

where V^* is the volume of the element after the deformation.

Dividing V^* by $V=dxdydz$ (i. e., the volume of the element before deformation) and denoting by Δ the relative change in volume due to the deformation, we have

$$1+\Delta = D = \begin{vmatrix} 1+\frac{\partial u}{\partial x}, & \frac{\partial u}{\partial y}, & \frac{\partial u}{\partial z} \\ \frac{\partial v}{\partial x}, & 1+\frac{\partial v}{\partial y}, & \frac{\partial v}{\partial z} \\ \frac{\partial w}{\partial x}, & \frac{\partial w}{\partial y}, & 1+\frac{\partial w}{\partial z} \end{vmatrix}. \tag{I.94}$$

To express Δ in terms of the strain components, let us square the result just obtained. Then, taking into account the rule for multiplying determinants, as well as (I.22), we get

$$(1+\Delta)^2 = \begin{vmatrix} 1+2\varepsilon_{xx}, & \varepsilon_{xy}, & \varepsilon_{xz} \\ \varepsilon_{xy}, & 1+2\varepsilon_{yy}, & \varepsilon_{yz} \\ \varepsilon_{xz}, & \varepsilon_{yz}, & 1+2\varepsilon_{zz} \end{vmatrix}. \qquad (I.95)$$

After expanding the determinant, this becomes

$$\begin{aligned}(1+\Delta)^2 &= (1+2\varepsilon_{xx})(1+2\varepsilon_{yy})(1+2\varepsilon_{zz})+2\varepsilon_{xy}\varepsilon_{xz}\varepsilon_{yz} \\ &\quad -(1+2\varepsilon_{xx})\,\varepsilon_{yz}^2-(1+2\varepsilon_{yy})\,\varepsilon_{xz}^2-(1+2\varepsilon_{zz})\,\varepsilon_{xy}^2 \\ &= 1+2a_2+4a_1+8a_0,\end{aligned} \qquad (I.96)$$

where a_0, a_1, a_2 are the invariants (I.87).

Finally, then,

$$\begin{aligned}\Delta &= \sqrt{1+2a_2+4a_1+8a_0}-1 \\ &= \sqrt{(1+2\varepsilon_1)(1+2\varepsilon_2)(1+2\varepsilon_3)}-1 \\ &= (1+E_1)(1+E_2)(1+E_3)-1,\end{aligned} \qquad (I.97)$$

where E_1, E_2, E_3 are the principal elongations at the point where the change in volume is calculated.

Note that the result obtained does not depend on the form of the element isolated from the body. This can be verified by examining a sphere of radius dS. Under the deformation, the given sphere becomes an ellipsoid with semiaxes $(1+E_1)\,dS$, $(1+E_2)\,dS$, $(1+E_3)\,dS$. If we subtract the volume of the sphere from the volume of this ellipsoid, and divide the difference by the volume of the element before the deformation (i. e., by the volume of the sphere), we have

$$\begin{aligned}\Delta &= \frac{\frac{4}{3}\pi\,[(1+E_1)(1+E_2)(1+E_3)-1]\,dS^3}{\frac{4}{3}\pi dS^3} \\ &= (1+E_1)(1+E_2)(1+E_3)-1.\end{aligned} \qquad (I.98)$$

This is identical with (I.97).

§ 12. On the Magnitude of Elongations and Shears

Up to now we have imposed no restrictions on the magnitude of the elongations and shears. Such generality in the theory of elasticity, however, is as a rule unnecessary. Only a small number of materials, such as rubber, for example, retain their elastic properties under large relative deformations. The vast majority of materials used in engineering (as, e. g., all metals and their alloys) are elastic only for elongations and shears which are very small (compared to unity).

As a typical example, let us examine steel, at present the most widely used material for building structures which are subjected to considerable loads. Fig. 11 shows the results of an experiment on the stretching of a sample of a brand of steel used in shipbuilding. As is customary, the relative elongations are laid off along the x-axis, and the stress along the y-axis.

In this graph, section OB corresponds to the elastic state of the performance of the steel, section BC, the plastic state (the region of flow),

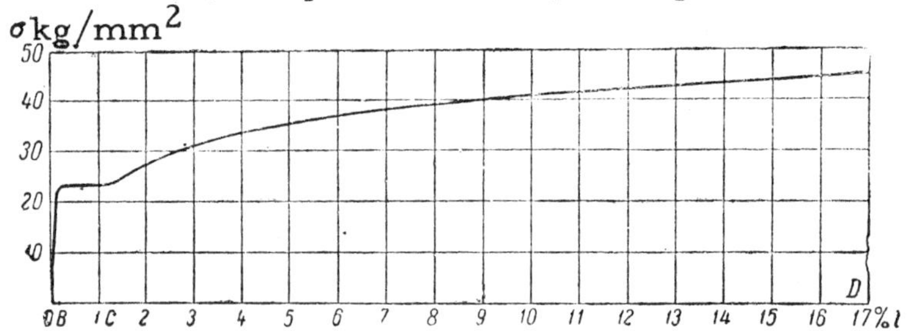

Fig. 11

and the greatest section, CD, the state of strengthening. The length of the elastic section in the given special case is equal to 1.15×10^{-3}, the length of the section of flow is 10^{-2}, and the length of the section of strengthening is 0.17. The length of the section of strengthening is actually still greater because of the formation of a neck on the sample. We shall not enter into these details here, however, as they are not required in the sequel.

These figures are sufficiently typical, and one can say that for steels in general, the range of elastic deformations is limited by relative elongations of the order of magnitude 10^{-3} to $5 \cdot 10^{-3}$. The limiting value of elastic shears for steel is of a similar order of magnitude.

For nonferrous metals (and their alloys) the figures are somewhat different. Their range of elastic deformation, however, is also limited by very small elongations and shears.

It follows that when the theory of elasticity is applied to metallic structures, it is natural and reasonable to simplify the formulas by neglecting the elongations and shears compared to unity. Thus, the greatest practical interest lies in the theory of small deformations, to which we now turn our attention.

§13. The Theory of Small Deformations

Let us simplify the formulas derived earlier, starting with the assumption that the elongations and shears are negligible compared to unity. Introducing this simplification into (I.28), we get

$$E_x = \varepsilon_{xx},\ E_y = \varepsilon_{yy},\ E_z = \varepsilon_{zz}. \tag{I.99}$$

Further, neglecting E_x, E_y, E_z, compared to unity, in formulas (I.32), and taking into consideration that the shears $\varphi_{xy}, \varphi_{xz}, \varphi_{yz}$ are small we obtain

$$\varphi_{xy} \approx \varepsilon_{xy},\ \varphi_{xz} \approx \varepsilon_{xz},\ \varphi_{yz} \approx \varepsilon_{yz}. \tag{I.100}$$

Thus, for small relative deformations, the components $\varepsilon_{xx}, \varepsilon_{yy}, \varepsilon_{zz}$ can be identified with the corresponding elongations, and the components ε_{xy}, $\varepsilon_{xz}, \varepsilon_{yz}$, with the corresponding shears.

Formula (I.97) for the increment of volume,

after neglecting the products of elongations compared to their first powers, assumes the form

$$\Delta \approx E_1 + E_2 + E_3, \tag{I.101}$$

or, if we take into account (I.99),

$$\Delta \approx \varepsilon_1 + \varepsilon_2 + \varepsilon_3 = a_2 = \varepsilon_{xx} + \varepsilon_{yy} + \varepsilon_{zz}. \tag{I.102}$$

It follows that, for small deformations, one can identify the invariant a_2 with the relative volumetric increment. Finally, if we make the analogous simplifications in Tables 1 and 2, we get Tables 5 and 6.

Table 5

	i_1	i_2	i_3
X	$1 + e_{xx}$	$\frac{1}{2} e_{xy} - \omega_z$	$\frac{1}{2} e_{xz} + \omega_y$
Y	$\frac{1}{2} e_{xy} + \omega_z$	$1 + e_{yy}$	$\frac{1}{2} e_{yz} - \omega_x$
Z	$\frac{1}{2} e_{xz} - \omega_y$	$\frac{1}{2} e_{yz} + \omega_x$	$1 + e_{zz}$

In the second of these, the magnitudes a_{ij} are determined from (I.15). The possibility of replacing Table 2 by Table 6 for small deformations follows from the fact that the parameters E_ξ, E_η, E_ζ are quantities of the same order of magnitude as the elongations, and that the determinant D, according to (I.94) and (I.98), is equal to

$$D = 1 + \Delta = (1 + E_1)(1 + E_2)(1 + E_3), \tag{I.103}$$

i. e., is equal to unity whenever the elongations can be neglected compared to unity.

It is clear from the above that the Tables 5 and 6 determine the cosines of the angles between $\mathbf{i}_1$, $\mathbf{i}_2$, $\mathbf{i}_3$ ($\mathbf{i}'_1$, $\mathbf{i}'_2$, $\mathbf{i}'_3$) and the X-, Y-, Z-axes to within an error which can be estimated by comparing the strain components with unity; with this degree of precision, the diagonal members of Tables 5 and 6 may exceed one.

Table 6

	$\mathbf{i}'_1$	$\mathbf{i}'_2$	$\mathbf{i}'_3$
X	α_{11}	α_{12}	α_{13}
Y	α_{21}	α_{22}	α_{23}
Z	α_{31}	α_{32}	α_{33}

We remark that if the rotations are large and the shears small, then the latter can be neglected in comparison with the former in determining the directions of the fibers in the strained state (this clearly does not imply that shears can generally be neglected in this case; the shears must be taken into account in those formulas where they occur alone without rotations). Examples of such a simplified approach to the determination of the directions of the fibers in the strained state are the hypothesis of plane sections in the theory of beams and Kirchhoff's hypothesis in the theory of plates. In either case the essence of the assumptions consists in neglecting the shears as compared to the rotations.

§ 14. The Case of Small Deformations and Small Angles of Rotation

If the angles of rotation as well as the strain components are small compared to unity, then the directions of the vectors $\mathbf{i}_1$, $\mathbf{i}_2$, $\mathbf{i}_3$ and $\mathbf{i}'_1$, $\mathbf{i}'_2$, $\mathbf{i}'_3$ will obviously deviate from those of X, Y, Z by only a small amount. As a result, the diagonal members of Tables 1 and 2 differ from unity only by quantities of the second order while the remaining members of these tables are quantities of the first order (if the maximum value of an angle of rotation is taken to be a quantity of the first order).

Let us examine the expression (I.77) for the mean value of the tangent of the angle of rotation of a volume element of the body about the Z-axis. Taking into account the last equation of (I.15), (I.77) reduces to the form

$$\overline{\operatorname{tg}\psi_z} = \frac{\omega_z}{\sqrt{\alpha_{33} - \omega_z^2}} = \frac{\omega_z}{\sqrt{\alpha_{33}}\sqrt{1 - \frac{\omega_z^2}{\alpha_{33}}}}, \tag{I.104}$$

whence

$$\overline{\sin\psi_z} = \frac{\omega_z}{\sqrt{\alpha_{33}}}. \tag{I.105}$$

In accordance with Table 6, however, $\alpha_{33} \approx \cos(i_3', Z)$ for a small deformation, and therefore

$$\overline{\sin\psi_z} \approx \frac{\omega_z}{\sqrt{\cos(i_3', Z)}}. \tag{I.106}$$

It was noted above that in the present case the cosine of the angle between the Z-axis and the vector $\mathbf{i}_3'$ differs from unity only by a quantity of the second order. Moreover, since the rotation is small, $\overline{\psi_z}$ differs from $\overline{\sin\psi_z}$ only by quantities of the third order. Hence, neglecting the squares of the angle of rotation compared to unity in formula (I.106), we obtain

$$\overline{\psi_z} \approx \omega_z. \tag{I.107}$$

Analogously, we also have

$$\overline{\psi_y} \approx \omega_y, \quad \overline{\psi_x} \approx \omega_x. \tag{I.107}$$

Thus, neglecting the squares of the angles of rotation compared to unity leads to the identification of the parameters ω_x, ω_y, ω_z with the mean rotations of a volume element about the X-, Y-, Z-axes. However, since small rotations are added vectorially, it can be said in this case that the volume element undergoes a rotation through an angle ω about the axis whose direction cosines are given in (I.62).

Let us furthermore examine how the formulas for the strain components may be simplified under the assumption that the angles of rotation and the strain components are small compared to unity.

The first of equations (I.22) can be written in the form

$$1+2\varepsilon_{xx}=(1+e_{xx})^2+\left(\frac{1}{2}\,e_{xy}+\omega_z\right)^2+\left(\frac{1}{2}e_{xz}-\omega_y\right)^2, \tag{I.108}$$

which is identically satisfied by putting

$$\frac{1+e_{xx}}{\sqrt{1+2\varepsilon_{xx}}}=\cos\varphi_1,\quad \frac{\frac{1}{2}\,e_{xy}+\omega_z}{\sqrt{1+2\varepsilon_{xx}}}=\sin\varphi_1\cos\chi_1,$$
$$\frac{\frac{1}{2}\,e_{xz}-\omega_y}{\sqrt{1+2\varepsilon_{xx}}}=\sin\varphi_1\sin\chi_1. \tag{I.109}$$

Similarly, the second and third of equations (I.22) are identically satisfied by the substitutions

$$\frac{\frac{1}{2}\,e_{xy}-\omega_z}{\sqrt{1+2\varepsilon_{yy}}}=\sin\varphi_2\sin\chi_2,\quad \frac{1+e_{yy}}{\sqrt{1+2\varepsilon_{yy}}}=\cos\varphi_2,$$
$$\frac{\frac{1}{2}\,e_{yz}+\omega_x}{\sqrt{1+2\varepsilon_{yy}}}=\sin\varphi_2\cos\chi_2 \tag{I.109}$$
$$\frac{\frac{1}{2}\,e_{xz}+\omega_y}{\sqrt{1+2\varepsilon_{zz}}}=\sin\varphi_3\cos\chi_3,\quad \frac{\frac{1}{2}\,e_{yz}-\omega_x}{\sqrt{1+2\varepsilon_{zz}}}=\sin\varphi_3\sin\chi_3,$$
$$\frac{1+e_{zz}}{\sqrt{1+2\varepsilon_{zz}}}=\cos\varphi_3.$$

Since, in accordance with Table I, φ_1, φ_2, and φ_3 are the angles between i_1 and the X-axis, i_2 and the Y-axis, i_3 and the Z-axis, respectively, and since these angles are of the same order of magnitude as the angles of rotation, it follows that they are small in this case. Consequently, we have

$$\frac{1+e_{xx}}{\sqrt{1+2\varepsilon_{xx}}}\approx 1-\frac{\varphi_1^2}{2},$$
$$\frac{1+e_{yy}}{\sqrt{1+2\varepsilon_{yy}}}\approx 1-\frac{\varphi_2^2}{2},$$
$$\frac{1+e_{zz}}{\sqrt{1+2\varepsilon_{zz}}}\approx 1-\frac{\varphi_3^2}{3}. \qquad \text{(I.110)}$$

Hence, since the strain components are assumed to be small, we obtain

$$\varepsilon_{xx}-e_{xx}\approx\frac{\varphi_1^2}{2},\quad \varepsilon_{yy}-e_{yy}\approx\frac{\varphi_2^2}{2},$$
$$\varepsilon_{zz}-e_{zz}\approx\frac{\varphi_3^2}{2}. \qquad \text{(I.111)}$$

Thus, in this case, the quantities e_{xx}, e_{yy}, e_{zz} differ from the corresponding strain components only by magnitudes of the same order as the squares of the angles of rotation.

Substituting (I.109) into the last three equations of (I.22), we obtain

$$\frac{\varepsilon_{xy}}{\sqrt{(1+2\varepsilon_{xx})(1+2\varepsilon_{yy})}}=\cos\varphi_1\sin\varphi_2\sin\chi_2$$
$$+\cos\varphi_2\sin\varphi_1\cos\chi_1+\sin\varphi_1\sin\varphi_2\sin\chi_1\cos\chi_2,$$
$$\frac{\varepsilon_{xz}}{\sqrt{(1+2\varepsilon_{xx})(1+2\varepsilon_{zz})}}=\cos\varphi_1\sin\varphi_3\cos\chi_3 \qquad \text{(I.112)}$$
$$+\cos\varphi_3\sin\varphi_1\sin\chi_1+\sin\varphi_1\sin\varphi_3\cos\chi_1\sin\chi_3,$$
$$\frac{\varepsilon_{yz}}{\sqrt{(1+2\varepsilon_{yy})(1+2\varepsilon_{zz})}}=\cos\varphi_2\sin\varphi_3\sin\gamma_3$$
$$+\cos\varphi_3\sin\varphi_2\cos\chi_2+\sin\varphi_2\sin\varphi_3\sin\chi_2\cos\chi_3.$$

Neglecting in (I.112) the strain components as

compared to unity and omitting all terms containing φ to higher than the second power, we arrive at

$$\begin{aligned} \varepsilon_{xy} &\approx \varphi_2 \sin\chi_2 + \varphi_1 \cos\chi_1 + \varphi_1\varphi_2 \sin\chi_1 \cos\chi_2, \\ \varepsilon_{xz} &\approx \varphi_3 \cos\chi_3 + \varphi_1 \sin\chi_1 + \varphi_1\varphi_3 \cos\chi_1 \sin\chi_3, \\ \varepsilon_{yz} &\approx \varphi_3 \sin\chi_3 + \varphi_2 \cos\chi_2 + \varphi_2\varphi_3 \sin\chi_2 \cos\chi_3. \end{aligned} \tag{I.113}$$

Introducing the analogous approximations in the equations (I.109) we find

$$\begin{aligned} &\frac{1}{2} e_{xy} + \omega_z \approx \varphi_1 \cos\chi_1, \quad \frac{1}{2} e_{xy} - \omega_z \approx \varphi_2 \sin\chi_2, \\ &\frac{1}{2} e_{xz} + \omega_y \approx \varphi_3 \cos\chi_3, \quad \frac{1}{2} e_{xz} - \omega_y \approx \varphi_1 \sin\chi_1, \\ &\frac{1}{2} e_{yz} + \omega_x \approx \varphi_2 \cos\chi_2, \quad \frac{1}{2} e_{yz} - \omega_x \approx \varphi_3 \sin\chi_3. \end{aligned} \tag{I.114}$$

As a result of these approximate expressions, equations (I.113) reduce to

$$\begin{aligned} \varepsilon_{xy} - e_{xy} &\approx \varphi_1\varphi_2 \sin\chi_1 \cos\chi_2, \\ \varepsilon_{xz} - e_{xz} &\approx \varphi_1\varphi_3 \cos\chi_1 \sin\chi_3, \\ \varepsilon_{yz} - e_{yz} &\approx \varphi_2\varphi_3 \sin\chi_2 \cos\chi_3, \end{aligned} \tag{I.115}$$

which imply that the parameters e_{xy}, e_{xz}, e_{yz} differ from the corresponding strain components only by quantities of the same order as the products of the angles of rotation.

These results enable us to simplify the formulas for the strain components on the assumption that the angles of rotation and the strain components can be neglected in comparison with unity.

It is seen that the squares of the parameters e_{xx}, e_{xy}, e_{xz} can be neglected in the first of equations (I.22),

$$\varepsilon_{xx} = e_{xx} + \frac{1}{2}\left[e_{xx}^2 + \left(\frac{1}{2} e_{xy} + \omega_z\right)^2 + \left(\frac{1}{2} e_{xz} - \omega_y\right)^2\right]. \tag{I.116}$$

This follows from the fact that taking these terms

into account, as was proved above, is equivalent to retaining the squares of the strain components and the fourth powers of the angles of rotation as well as the first powers of the strain components and the squares of the angles of rotation. This would be contrary to the method of approximation adopted.

Omitting the terms indicated, we arrive at the equation

$$\varepsilon_{xx} = e_{xx} + \frac{1}{2}\left[e_{xy}\,\omega_z - e_{xz}\,\omega_y + \omega_y^2 + \omega_z^2\right]. \quad \text{(I.117)}$$

Substitution in (I.117) for e_{xy} and e_{xz} of their expressions in terms of the strain components and the angles of rotation introduces the squares of the strain components and the cubes of the angles of rotation. This implies that the first two terms in the square bracket of equation (I.117) are substantially smaller than the last two terms in the same bracket, and enables us to write

$$\varepsilon_{xx} \approx e_{xx} + \frac{1}{2}(\omega_y^2 + \omega_z^2). \quad \text{(I.118)}$$

The remaining equations of (I.22) can be simplified in an analogous fashion. The final approximate expressions for the strain components are

$$\begin{aligned} &\varepsilon_{xx} \approx e_{xx} + \frac{1}{2}(\omega_y^2 + \omega_z^2), \quad \varepsilon_{xy} \approx e_{xy} - \omega_x\omega_y, \\ &\varepsilon_{yy} \approx e_{yy} + \frac{1}{2}(\omega_x^2 + \omega_z^2), \quad \varepsilon_{xz} \approx e_{xz} - \omega_x\omega_z, \quad \text{(I.119)} \\ &\varepsilon_{zz} \approx e_{zz} + \frac{1}{2}(\omega_x^2 + \omega_y^2), \quad \varepsilon_{yz} \approx e_{yz} - \omega_y\omega_z. \end{aligned}$$

These are correct to within the accuracy obtainable by neglecting the angles of rotation and the strain components compared to unity.

It must be noted that these formulas may be further simplified in special cases. This may happen, for example, if one or two of the components of an angle of rotation are extremely small. They can then be neglected, and only the essential component (or components) need be retained.

In conclusion, let us emphasize as an important point of the above discussion that the smallness of the rotations and strains in comparison with unity does not at all imply that they are of the same order of magnitude. It is precisely for this reason that it is necessary to retain, in equations (I.119), the terms which contain rotations to the second power as well as the terms containing the parameters $e_{xx}, \ldots, e_{yz}$ to the first power

§ 15. The Transition to the Equations of the Classical Theory

Let us now assume that the squares and products of the angles of rotation may be neglected in equations (I.119). This assumption leads to the following expressions for the strain components:

$$\begin{aligned} \varepsilon_{xx} &\approx e_{xx} = \frac{\partial u}{\partial x}, & \varepsilon_{xy} &\approx e_{xy} = \frac{\partial u}{\partial y} + \frac{\partial v}{\partial x}, \\ \varepsilon_{yy} &\approx e_{yy} = \frac{\partial v}{\partial y}, & \varepsilon_{xz} &\approx e_{xz} = \frac{\partial u}{\partial z} + \frac{\partial w}{\partial x}, \\ \varepsilon_{zz} &\approx e_{zz} = \frac{\partial w}{\partial z}, & \varepsilon_{yz} &\approx e_{yz} = \frac{\partial v}{\partial z} + \frac{\partial w}{\partial y}. \end{aligned} \quad (I.120)$$

These are the equations of the classical theory of elasticity.

It is seen from the two preceding sections that the expressions for the strain components become linear only under the two following conditions:

a) the elongations, shears, and angles of rotation must be small compared to unity.

b) the terms of the second degree in the angles

of rotation appearing in (I.119) must be small compared to the corresponding strain components.

The last requirement can be formulated, *roughly speaking*, as the condition that the squares of the angles of rotation be negligibly small compared to the elongations and shears. If the body is *massive*, i. e., is of the same order of magnitude in all three of its dimensions, then condition a) implies condition b). This is not true if the body is *flexible*, i. e., if its extension in one or two directions is essentially small compared to its remaining dimensions (rod, plate, shell). In this case the angles of rotation may considerably exceed the elongations and shears, so that equations (I.120) are in general not applicable to such bodies. This implies that the linear equations (I.120) are to be used primarily in analyzing the deformation of massive bodies, while the nonlinear equations (I.22) and (I.119) are applicable to the deformation of flexible bodies.

Equations (I.120) are however also applicable to some problems dealing with the deformation of flexible bodies. In particular, they may be used in the study of the bending and torsion of thin rods (under the conditions that the rod is not subjected to the simultaneous action of axial forces and that the angles of rotation of its sections, while large compared to the strains, are small compared to unity).

It is not difficult to see, for example, why torsion remains a problem of the classical theory. In this problem only one of the angles of rotation (ω_x, if the X-axis is directed along the rod) is large compared to the shears, while the other two angles of rotation are of the same order of magnitude as the strains. At the same time only the two strain components

$$\varepsilon_{xy} \approx e_{xy} - \omega_x \omega_y,$$
$$\varepsilon_{xz} \approx e_{xz} - \omega_x \omega_z,$$

play an essential role in torsion. Here the last terms in the two expressions can be ignored (in spite of the fact that ω_x significantly exceeds e_{xy} and e_{xz}) since, as remarked above, ω_y and ω_z are very small. This example indicates that it is not always correct to delimit the domain of applicability of equations (I.120) by the requirement that the squares of the angles of rotation be negligibly small compared to the strains. Actually, the domain of applicability of the linear equations is somewhat broader.

In conclusion, we must call special attention to the tradition, established in the majority of books on the theory of elasticity, of referring to the expressions (I.22) as the "components of a finite deformation". This inevitably implies (although this is not explicitly stated) that the expressions (I.120) of the classical theory are the "components of an infinitesimal deformation". This chapter makes it completely clear, however, that the degree of smallness of the elongations and shears compared to unity is not at all a sufficient criterion for passing from equations (I.22) or (I.119) to equations (I.120). The magnitude of the angles of rotation also plays an essential role in the solution of this question.

In some problems the use of the linear equations is inadmissible even for very small elongations and shears (compression of a thin rod, bending of a thin plate). In other problems the linear equations are applicable even though the elongations and shears are much larger (extension of a rod, bending of a thick plate).

Thus, both the nonlinear and the classical theory of elasticity deal with finite deformations, and, moreover, as a rule, with deformations of the same order of smallness. Otherwise, the classical theory would have no practical significance at all.

The difference in approach of these two theories in dealing with the determination of strain consists only in that the linear theory neglects the influence of rotations on elongations and shears, while the nonlinear theory takes it into account. As a result, the nonlinear theory embraces all problems dealing with the elastic deformation of bodies, while the linear theory applies only to a particular circle of problems. This does not mean that we wish to minimize the significance of the classical theory, which yields solutions, completely in agreement with experience, to many problems of practical importance. In such cases, there is of course no need to revert to the nonlinear theory. However, by no means all problems are embraced by the classical theory. The latter excludes a whole series of problems (see the Introduction), which must therefore be examined in the spirit of the nonlinear theory.

The remarks above make it sufficiently clear why we cannot call the nonlinear theory of deformations the theory of finite deformations.

§ 16. On the Transition to Curvilinear Coordinates

It has been assumed up to now that the positions of the points of a body are expressed in terms of Cartesian coordinates X, Y, Z. In the solution of some concrete problems it is more convenient to use curvilinear coordinates. In this connection, let us develop the rules of transformation of the equations derived above, to arbitrary orthogonal curvilinear coordinates.

Let the curvilinear coordinates be related to the Cartesian coordinates in accordance with the equations

$$x = f_1(\alpha_1, \alpha_2, \alpha_3), \; y = f_2(\alpha_1, \alpha_2, \alpha_3), \; z = f_3(\alpha_1, \alpha_2, \alpha_3). \quad \text{(I.121)}$$

These equations determine three families of curves, the coordinate lines α_1, α_2, α_3, with the property that each point of space appears as an intersection of three lines belonging to different families.

Denote the unit vectors tangent to the coordinate lines by $\mathbf{k}_1$, $\mathbf{k}_2$, $\mathbf{k}_3$, respectively. Since the curvilinear coordinates are assumed to be orthogonal, these vectors will form at every point a mutually perpendicular trihedral of local coordinate axes (we refer to them as local axes because, unlike a Cartesian system, the direction of these changes from one point to another). Let us denote by u, v, w the projections of the displacement of an arbitrary point of the body on the directions $\mathbf{k}_1$, $\mathbf{k}_2$, $\mathbf{k}_3$ at the given point.

It is proved in books on the linear theory of elasticity (see, for example, Leibenson, 1, or Love, 1) that to the parameters e_{xx}, e_{yy}, e_{zz}, e_{xy}, e_{xz}, e_{yz} there correspond, in orthogonal curvilinear coordinates, the expressions

$$\begin{aligned} e_{11} &= \frac{1}{H_1}\frac{\partial u}{\partial \alpha_1} + \frac{1}{H_1 H_2}\frac{\partial H_1}{\partial \alpha_2} v + \frac{1}{H_1 H_3}\frac{\partial H_1}{\partial \alpha_3} w, \\ e_{22} &= \frac{1}{H_2}\frac{\partial v}{\partial \alpha_2} + \frac{1}{H_2 H_3}\frac{\partial H_2}{\partial \alpha_3} w + \frac{1}{H_1 H_2}\frac{\partial H_2}{\partial \alpha_1} u, \\ e_{33} &= \frac{1}{H_3}\frac{\partial w}{\partial \alpha_3} + \frac{1}{H_1 H_3}\frac{\partial H_3}{\partial \alpha_1} u + \frac{1}{H_2 H_3}\frac{\partial H_3}{\partial \alpha_2} v, \end{aligned} \quad \text{(I.122)}$$

$$e_{12}=\frac{H_2}{H_1}\frac{\partial}{\partial\alpha_1}\left(\frac{v}{H_2}\right)+\frac{H_1}{H_2}\frac{\partial}{\partial\alpha_2}\left(\frac{u}{H_1}\right),$$
$$e_{13}=\frac{H_1}{H_3}\frac{\partial}{\partial\alpha_3}\left(\frac{u}{H_1}\right)+\frac{H_3}{H_1}\frac{\partial}{\partial\alpha_1}\left(\frac{w}{H_3}\right),$$
$$e_{23}=\frac{H_3}{H_2}\frac{\partial}{\partial\alpha_2}\left(\frac{w}{H_3}\right)+\frac{H_2}{H_3}\frac{\partial}{\partial\alpha_3}\left(\frac{v}{H_2}\right),$$

where H_1, H_2, H_3 are the Lamé coefficients

$$H_1=\sqrt{\left(\frac{\partial x}{\partial\alpha_1}\right)^2+\left(\frac{\partial y}{\partial\alpha_1}\right)^2+\left(\frac{\partial z}{\partial\alpha_1}\right)^2},$$
$$H_2=\sqrt{\left(\frac{\partial x}{\partial\alpha_2}\right)^2+\left(\frac{\partial y}{\partial\alpha_2}\right)^2+\left(\frac{\partial z}{\partial\alpha_2}\right)^2}, \qquad (I.123)$$
$$H_3=\sqrt{\left(\frac{\partial x}{\partial\alpha_3}\right)^2+\left(\frac{\partial y}{\partial\alpha_3}\right)^2+\left(\frac{\partial z}{\partial\alpha_3}\right)^2}.$$

Similarly, the parameters ω_x, ω_y, ω_z, referred to arbitrary orthogonal curvilinear coordinates, are given by the equations

$$2\omega_1=\frac{1}{H_2H_3}\left[\frac{\partial}{\partial\alpha_2}(H_3w)-\frac{\partial}{\partial\alpha_3}(H_2v)\right],$$
$$2\omega_2=\frac{1}{H_1H_3}\left[\frac{\partial}{\partial\alpha_3}(H_1u)-\frac{\partial}{\partial\alpha_1}(H_3w)\right], \qquad (I.124)$$
$$2\omega_3=\frac{1}{H_1H_2}\left[\frac{\partial}{\partial\alpha_1}(H_2v)-\frac{\partial}{\partial\alpha_2}(H_1u)\right],$$

which are the well-known expressions for the components of a curl.

In the linear theory, based on the assumption that the strains and rotations are small and of comparable magnitude, the parameters e_{11}, e_{22}, e_{33} are the elongations of the line elements having directions $\mathbf{k}_1$, $\mathbf{k}_2$, $\mathbf{k}_3$, the parameters e_{12}, e_{13}, e_{23} are shears between these elements, and ω_1, ω_2, ω_3 are the angles of rotation of a volume element about $\mathbf{k}_1$ $\mathbf{k}_2$, $\mathbf{k}_3$. In the general case of an arbitrary deformation these parameters do not have so simple a geometric meaning and cannot be i-dentified with the strain components and the angles of rotation. Equations (I.22) are the expressions for the strain components in a Cartesian coordinate system. These equations contain the parameters e and ω, whose rules of transformation have already been derived. This enables us to write down immediately the following expressions for the strain components referred to an arbitrary orthogonal coordinate system:

$$
\begin{aligned}
\varepsilon_{11} &= e_{11} + \frac{1}{2}\left[e_{11}^2 + \left(\frac{1}{2}e_{12} + \omega_3\right)^2 + \left(\frac{1}{2}e_{13} - \omega_2\right)^2\right], \\
\varepsilon_{22} &= e_{22} + \frac{1}{2}\left[e_{22}^2 + \left(\frac{1}{2}e_{12} - \omega_3\right)^2 + \left(\frac{1}{2}e_{23} + \omega_1\right)^2\right], \\
\varepsilon_{33} &= e_{33} + \frac{1}{2}\left[e_{33}^2 + \left(\frac{1}{2}e_{13} + \omega_2\right)^2 + \left(\frac{1}{2}e_{23} - \omega_1\right)^2\right], \\
\varepsilon_{12} &= e_{12} + e_{11}\left(\frac{1}{2}e_{12} - \omega_3\right) + e_{22}\left(\frac{1}{2}e_{12} + \omega_3\right) \\
&\quad + \left(\frac{1}{2}e_{13} - \omega_2\right)\left(\frac{1}{2}e_{23} + \omega_1\right), \qquad \text{(I.125)} \\
\varepsilon_{13} &= e_{13} + e_{11}\left(\frac{1}{2}e_{13} + \omega_2\right) + e_{33}\left(\frac{1}{2}e_{13} - \omega_2\right) \\
&\quad + \left(\frac{1}{2}e_{12} + \omega_3\right)\left(\frac{1}{2}e_{23} - \omega_1\right), \\
\varepsilon_{23} &= e_{23} + e_{22}\left(\frac{1}{2}e_{23} - \omega_1\right) + e_{33}\left(\frac{1}{2}e_{23} + \omega_1\right) \\
&\quad + \left(\frac{1}{2}e_{12} - \omega_3\right)\left(\frac{1}{2}e_{13} + \omega_2\right).
\end{aligned}
$$

Here the parameters e and ω are determined by (I.122) and (I.124).

The relative elongations of the line elements having the directions $\mathbf{k}_1$, $\mathbf{k}_2$, $\mathbf{k}_3$ in the unstrained state are

$$
\begin{aligned}
E_1 &= \sqrt{1 + 2\varepsilon_{11}} - 1, \\
E_2 &= \sqrt{1 + 2\varepsilon_{22}} - 1, \\
E_3 &= \sqrt{1 + 2\varepsilon_{33}} - 1.
\end{aligned}
$$

Correspondingly, the shears between these line elements are given by

$$\sin\varphi_{12} = \frac{\varepsilon_{12}}{(1+E_1)(1+E_2)},$$
$$\sin\varphi_{13} = \frac{\varepsilon_{13}}{(1+E_1)(1+E_3)}, \qquad (I.126)$$
$$\sin\varphi_{23} = \frac{\varepsilon_{23}}{(1+E_2)(1+E_3)}.$$

If the elongations and shears are small, these equations can be simplified in a fashion similar to that of § 13. Accordingly, these formulas may also be simplified in other ways discussed in the preceding sections.

CHAPTER II

THE EQUILIBRIUM OF AN ELEMENT OF VOLUME OF A BODY

§ 17. Stresses

In this chapter we shall investigate the conditions for the equilibrium of an arbitrary infinitesimal element of volume of the deformed body.

If we isolate such an element, it is necessary to apply to it forces distributed over its bounding surface which represent the effect of the surrounding medium on this element. Consider an element of area $d\Omega$ on the given surface. Its orientation can be described by a unit vector $\mathbf{n}$ along the normal, which will be regarded as positive if directed toward the exterior of the element of volume in question. Denote by $\boldsymbol{\sigma} d\Omega$ the force acting on the element of area. The vector $\boldsymbol{\sigma}$ represents the intensity of the surface loading on the area $d\Omega$. Its magnitude and direction depend on the position of the area (which can be specified by the coordinates ξ, η, ζ of its centroid) as well as on the orientation of the area (i. e., on $\mathbf{n}$). The triple ξ, η, ζ, however, determines a radius vector $\mathbf{r}$ extending from the origin of coordinates to the centroid of the area, so that

$$\boldsymbol{\sigma} = \boldsymbol{\sigma}(\mathbf{r}, \mathbf{n}). \tag{II.1}$$

Thus, the intensity of the surface loading is a function of two vectors. This function is odd with respect to $\mathfrak{n}$:

$$\sigma(\mathfrak{r}, \mathfrak{n}) = -\sigma(\mathfrak{r}, -\mathfrak{n}). \tag{II.2}$$

This is due to the fact that the action of the medium on the area $d\Omega$ of the element of volume must be equal in magnitude and opposite in direction to the action of this area on the contiguous medium, whose orientation is given by the unit vector $-\mathfrak{n}$.

The vector σ is called the stress. In the sequel it will be marked with a subscript indicating the direction of the normal to the area on which it acts.

It suffices to know the values of the stresses on three mutually perpendicular areas passing through the point M^* of the deformed body in order to be able to find the stress on any other area passing through the same point. To show this, let us examine an element of volume which is a tetrahedron, three of whose edges are parallel to the coordinate axes X, Y, Z and equal $d\xi$, $d\eta$, $d\zeta$, respectively.

For the given element to be in equilibrium, it is necessary, first of all, that the sum of all the forces acting on it be equal to zero. Bear in mind that, in addition to the surface forces, there are mass forces exerted on the element, which act throughout the body (e. g., gravitational or inertial forces) and which can be considered as constant from point to point in view of the smallness of the element.

In accordance with this remark, the vanishing of the sum of all the forces acting on the tetrahedron can be written as follows:

$$\sigma_n d\Omega + \sigma_{-\xi}\frac{d\eta d\zeta}{2} + \sigma_{-\eta}\frac{d\xi d\zeta}{2} + \sigma_{-\zeta}\frac{d\xi d\eta}{2} + F\frac{d\xi d\eta d\zeta}{6} = 0. \tag{II.3}$$

Here

$d\Omega$ is the area of the inclined face of the tetrahedron;

$\mathbf{n}$ is the unit vector of its external normal;

F is the mean value of the specific body force acting on the tetrahedron (e. g., the mean value of the specific weight or the mean value of the inertial forces referred to the unit volume;

$\sigma_{-\xi}$ is the stress on the area, perpendicular to the X-axis;

$\sigma_{-\eta}$ is the stress on the area, perpendicular to the Y-axis;

$\sigma_{-\zeta}$ is the stress on the area, perpendicular to the Z-axis.

The subscripts of the last three stresses are negative because the directions of the external normals to the corresponding areas are opposite to those of the coordinate axes (Fig. 12).

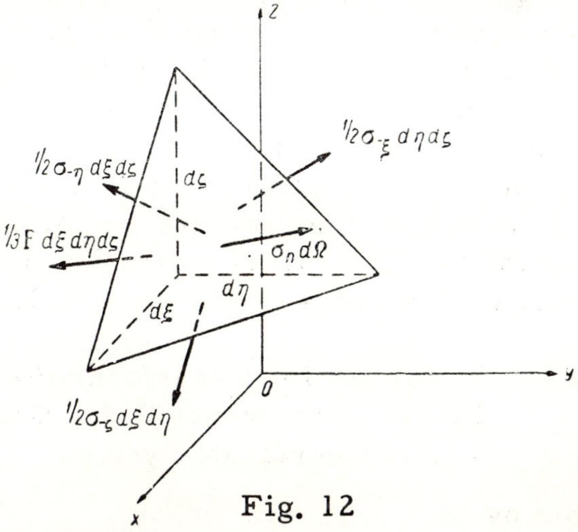

Fig. 12

If we divide (II.3) by $d\Omega$ and note (II.2) we obtain

$$\sigma_n = \sigma_\xi \frac{d\eta d\zeta}{2d\Omega} + \sigma_\eta \frac{d\xi d\zeta}{2d\Omega} + \sigma_\zeta \frac{d\xi d\eta}{2d\Omega} - F\frac{d\xi d\eta d\zeta}{6d\Omega}, \quad \text{(II.4)}$$

where $\frac{d\eta d\zeta}{2}$, $\frac{d\xi d\zeta}{2}$, $\frac{d\xi d\eta}{2}$ are the projections of the inclined face of the tetrahedron on the YZ-, XZ-, XY-planes, so that

$$\frac{d\eta d\zeta}{2d\Omega} = \cos(\mathrm{n}, X), \quad \frac{d\xi d\zeta}{2d\Omega} = \cos(\mathrm{n}, Y), \quad \frac{d\xi d\eta}{2d\Omega} = \cos(\mathrm{n}, Z). \quad \text{(II.5)}$$

Furthermore, the fraction $\frac{d\xi d\eta d\zeta}{6d\Omega}$ represents the ratio of the volume of the tetrahedron to the area of its inclined face, and is therefore a magnitude of the order of the linear dimension of the tetrahedron (i. e., an infinitesimal quantity). Hence, the last term in (II.4) is also an infinitesimal, and can be neglected, because the remaining terms are finite. We thus arrive at the vector equation

$$\sigma_n = \sigma_\xi \cos(\mathrm{n}, X) + \sigma_\eta \cos(\mathrm{n}, Y) + \sigma_\zeta \cos(\mathrm{n}, Z), \quad \text{(II.6)}$$

which expresses the stress at the point M^*, on the area whose external normal is given by the unit vector n, in terms of the three stresses on the areas perpendicular to the coordinate axes. The state of stress of the body in the neighborhood of an arbitrary point M^* is thus completely characterized by the vectors σ_ξ, σ_η, σ_ζ whose magnitudes and directions evidently depend on the choice of directions of the axes X, Y, Z.

§ 18. Formulas for Transformation of Stress Components Under Change of Coordinate System

Denote by

$\sigma_{\xi\xi}$, $\sigma_{\xi\eta}$, $\sigma_{\xi\zeta}$ the projections of σ_ξ on X, Y, Z,
$\sigma_{\eta\xi}$, $\sigma_{\eta\eta}$, $\sigma_{\eta\zeta}$ the projections of σ_η on X, Y, Z,
$\sigma_{\zeta\xi}$, $\sigma_{\zeta\eta}$, $\sigma_{\zeta\zeta}$ the projections of σ_ζ on X, Y, Z.

We call the nine quantities σ_{ij} the components of the state of stress in the system X, Y, Z.

Let us see how these quantities are expressed if, instead of the system X, Y, Z, we take another rectangular system X', Y', Z', the directions of whose axes relative to the axes of the first system are given by the cosines in Table 7.

Table 7

	X	Y	Z
X'	l_1	m_1	n_1
Y'	l_2	m_2	n_2
Z'	l_3	m_3	n_3

Clearly l_1, m_1, n_1 are the cosines of the normal to the area parallel to the Z' Y'-plane (since X is normal to this plane). Hence, according to (II.6),

$$\sigma_{\xi'} = \sigma_\xi l_1 + \sigma_\eta m_1 + \sigma_\zeta n_1. \tag{II.7}$$

If we project this expression on the X'-, Y'-, Z'-axes, we get

$$\begin{aligned} \sigma_{\xi'\xi'} &= (\sigma_\xi)_{\xi'} l_1 + (\sigma_\eta)_{\xi'} m_1 + (\sigma_\zeta)_{\xi'} n_1, \\ \sigma_{\xi'\eta'} &= (\sigma_\xi)_{\eta'} l_1 + (\sigma_\eta)_{\eta'} m_1 + (\sigma_\zeta)_{\eta'} n_1, \\ \sigma_{\xi'\zeta'} &= (\sigma_\xi)_{\zeta'} l_1 + (\sigma_\eta)_{\zeta'} m_1 + (\sigma_\zeta)_{\zeta'} n_1, \end{aligned} \tag{II.8}$$

and the projections of the vectors σ_ξ, σ_η, σ_ζ on the X' -, Y' -, Z' -axes can be expressed in terms of their projections on the X-, Y-, Z-axes as follows:

$$\begin{aligned} (\sigma_\xi)_{\xi'} &= \sigma_{\xi\xi} \cos(X, X') + \sigma_{\xi\eta} \cos(Y, X') \\ &\quad + \sigma_{\xi\zeta} \cos(Z, X') = \sigma_{\xi\xi} l_1 + \sigma_{\xi\eta} m_1 + \sigma_{\xi\zeta} n_1, \\ (\sigma_\eta)_{\xi'} &= \sigma_{\eta\xi} l_1 + \sigma_{\eta\eta} m_1 + \sigma_{\eta\zeta} n_1, \\ (\sigma_\zeta)_{\xi'} &= \sigma_{\zeta\xi} l_1 + \sigma_{\zeta\eta} m_1 + \sigma_{\zeta\zeta} n_1, \\ (\sigma_\xi)_{\eta'} &= \sigma_{\xi\xi} l_2 + \sigma_{\xi\eta} m_2 + \sigma_{\xi\zeta} n_2, \\ (\sigma_\eta)_{\eta'} &= \sigma_{\eta\xi} l_2 + \sigma_{\eta\eta} m_2 + \sigma_{\eta\zeta} n_2, \\ (\sigma_\zeta)_{\eta'} &= \sigma_{\zeta\xi} l_2 + \sigma_{\zeta\eta} m_2 + \sigma_{\zeta\zeta} n_3 \\ (\sigma_\xi)_{\zeta'} &= \sigma_{\xi\xi} l_3 + \sigma_{\xi\eta} m_3 + \sigma_{\xi\zeta} n_3, \\ (\sigma_\eta)_{\zeta'} &= \sigma_{\eta\xi} l_3 + \sigma_{\eta\eta} m_3 + \sigma_{\eta\zeta} n_3, \\ (\sigma_\zeta)_{\zeta'} &= \sigma_{\zeta\xi} l_3 + \sigma_{\zeta\eta} m_3 + \sigma_{\zeta\zeta} n_3. \end{aligned} \tag{II.9}$$

Substituting these expressions in (II.8) we obtain

$$\begin{aligned}\sigma'_{\xi\xi} = \sigma_{\xi'\xi'} &= \sigma_{\xi\xi}\, l_1^2 + \sigma_{\eta\eta}\, m_1^2 + \sigma_{\zeta\zeta}\, n_1^2 + (\sigma_{\xi\eta} + \sigma_{\eta\xi})\, m_1 l_1 \\ &+ (\sigma_{\xi\zeta} + \sigma_{\zeta\xi}) l_1 n_1 + (\sigma_{\eta\zeta} + \sigma_{\zeta\eta})\, m_1 n_1, \\ \sigma'_{\xi\eta} = \sigma_{\xi'\eta'} &= \sigma_{\xi\xi}\, l_1 l_2 + \sigma_{\eta\eta}\, m_1 m_2 + \sigma_{\zeta\zeta}\, n_1 n_2 \\ &+ \sigma_{\xi\eta}\, l_1 m_2 + \sigma_{\eta\xi}\, l_2 m_1 + \sigma_{\xi\zeta}\, l_1 n_2 + \sigma_{\zeta\xi}\, l_2 n_1 \\ &+ \sigma_{\eta\zeta}\, m_1 n_2 + \sigma_{\zeta\eta}\, n_1 m_2, \\ \sigma'_{\xi\zeta} = \sigma_{\xi'\zeta'} &= \sigma_{\xi\xi} l_1 l_3 + \sigma_{\eta\eta}\, m_1 m_3 + \sigma_{\zeta\zeta}\, l_1 m_3 + \sigma_{\eta\xi}\, l_3 m_1 \\ &+ \sigma_{\xi\zeta}\, l_1 n_3 + \sigma_{\zeta\xi}\, l_3 n_1 + \sigma_{\eta\zeta}\, m_1 n_3 + \sigma_{\zeta\eta}\, n_1 m_3.\end{aligned} \tag{II.10}$$

The expressions for the remaining stress components can be written down analogously. In any coordinate system, as we shall show in a subsequent section,

$$\sigma_{ij} = \sigma_{ji}, \tag{II.11}$$

i. e., the magnitudes of the stress components do not change if the order of the subscripts is reversed.

If we take into account this fact, which will be proved later, we have, finally,

$$\begin{aligned}\sigma'_{\xi\xi} &= \sigma_{\xi\xi}\, l_1^2 + \sigma_{\eta\eta}\, m_1^2 + \sigma_{\zeta\zeta}\, n_1^2 + 2\sigma_{\xi\eta}\, l_1 m_1 + 2\sigma_{\xi\zeta}\, l_1 n_1 + 2\sigma_{\eta\zeta}\, m_1 n_1, \\ \sigma'_{\eta\eta} &= \sigma_{\xi\xi}\, l_2^2 + \sigma_{\eta\eta}\, m_2^2 + \sigma_{\zeta\zeta}\, n_2^2 + 2\sigma_{\xi\eta}\, l_2 m_2 + 2\sigma_{\xi\zeta} l_2 n_2 \\ &\quad + 2\sigma_{\eta\zeta}\, m_2 n_2, \\ \sigma'_{\zeta\zeta} &= \sigma_{\xi\xi}\, l_3^2 + \sigma_{\eta\eta}\, m_3^2 + \sigma_{\zeta\zeta}\, n_3^2 + 2\sigma_{\xi\eta}\, l_3 m_3 + 2\sigma_{\xi\zeta}\, l_3 n_3 + 2\sigma_{\eta\zeta}\, m_3 n_3, \\ \sigma'_{\xi\eta} &= \sigma'_{\eta\xi} = \sigma_{\xi\xi}\, l_1 l_2 + \sigma_{\eta\eta}\, m_1 m_2 + \sigma_{\zeta\zeta}\, n_1 n_2 + \sigma_{\xi\eta}\, (l_1 m_2 + l_2 m_1) \\ &\quad + \sigma_{\xi\zeta}\, (l_1 n_2 + l_2 n_1) + \sigma_{\eta\zeta}\, (m_1 n_2 + m_2 n_1), \\ \sigma'_{\zeta\xi} &= \sigma'_{\xi\zeta} = \sigma_{\xi\xi}\, l_1 l_3 + \sigma_{\eta\eta}\, m_1 m_3 + \sigma_{\zeta\zeta}\, n_1 n_3 + \sigma_{\xi\eta}\, (l_1 m_3 + l_3 m_1) \\ &\quad + \sigma_{\xi\zeta}\, (l_1 n_3 + l_3 n_1) + \sigma_{\eta\zeta}\, (m_1 n_3 + m_3 n_1), \\ \sigma'_{\eta\zeta} &= \sigma'_{\zeta\eta} = \sigma_{\xi\xi}\, l_2 l_3 + \sigma_{\eta\eta}\, m_2 m_3 + \sigma_{\zeta\zeta}\, n_2 n_3 + \sigma_{\xi\eta}\, (l_2 m_3 + l_3 m_2) \\ &\quad + \sigma_{\xi\zeta}\, (l_2 n_3 + l_3 n_2) + \sigma_{\eta\zeta}\, (m_2 n_3 + m_3 n_2).\end{aligned} \tag{II.12}$$

Comparing (II.12) with (I.36), we see that the transformation of the stress components under a change of axes is similar to that of the quantities

$$\left.\begin{matrix} \varepsilon_{xx}, & \frac{1}{2}\varepsilon_{xy}, & \frac{1}{2}\varepsilon_{xz}, \\ \frac{1}{2}\varepsilon_{yx}, & \varepsilon_{yy}, & \frac{1}{2}\varepsilon_{yz} \\ \frac{1}{2}\varepsilon_{zx}, & \frac{1}{2}\varepsilon_{zy}, & \varepsilon_{zz} \end{matrix}\right\}$$

For this reason, the series of results proved in the preceding chapter for the strain components can immediately be asserted also for the stress components. In particular, we can say that at every point of the deformed body there exist three mutually perpendicular directions for which $\sigma'_{\xi\eta}$, $\sigma'_{\xi\zeta}$, $\sigma'_{\eta\zeta}$ are equal to zero.

The cosines l, m, n determining one of these directions can be found from the equations

$$\begin{aligned} (\sigma_{\xi\xi} - \sigma)\, l + \sigma_{\xi\eta}\, m + \sigma_{\xi\zeta}\, n &= 0, \\ \sigma_{\xi\eta}\, l + (\sigma_{\eta\eta} - \sigma)\, m + \sigma_{\eta\zeta}\, n &= 0, \\ \sigma_{\xi\zeta}\, l + \sigma_{\eta\zeta}\, m + (\sigma_{\zeta\zeta} - \sigma)\, n &= 0 \end{aligned} \tag{II.13}$$

and the relation

$$l^2 + m^2 + n^2 = 1, \tag{II.14}$$

where the parameter σ which appears in the system (II.13) is one of the three roots of the cubic equation

$$\begin{vmatrix} \sigma_{\xi\xi} - \sigma, & \sigma_{\xi\eta}, & \sigma_{\xi\zeta} \\ \sigma_{\xi\eta}, & \sigma_{\eta\eta} - \sigma, & \sigma_{\eta\zeta} \\ \sigma_{\xi\zeta}, & \sigma_{\eta\zeta}, & \sigma_{\zeta\zeta} - \sigma \end{vmatrix} = 0. \tag{II.15}$$

Since the elements of this determinant are symmetric with respect to its main diagonal, all the roots of (II.15) are real. Hence one can write down three systems of the form (II.13) from which one can find all three directions corresponding to the extremal values of $\sigma_{\xi\xi}$, $\sigma_{\eta\eta}$, $\sigma_{\zeta\zeta}$ and the zero values of $\sigma_{\xi\eta}$, $\sigma_{\xi\zeta}$, $\sigma_{\eta\zeta}$. The extremal values of $\sigma_{\xi\xi}$, $\sigma_{\eta\eta}$, $\sigma_{\zeta\zeta}$ (to be denoted in the sequel by σ_1, σ_2, σ_3) are equal to the roots of (II.15).

We agree to call $\sigma_{\xi\xi}$, $\sigma_{\eta\eta}$, $\sigma_{\zeta\zeta}$ the normal stresses on areas perpendicular to the coordinate axes, and $\sigma_{\xi\eta}$, $\sigma_{\xi\zeta}$, $\sigma_{\eta\zeta}$, the tangential stresses on the same areas.

Accordingly, σ_1, σ_2, σ_3, the extremal values of the normal stresses at the point M^*, will be called the principal normal stresses at this point, whereas the directions of the normals to the areas on which they act are called the principal axes of the state of stress.

Formulas (II.13), (II.14), and (II.15) enable one to determine at every point of the body the magnitudes of the principal stresses and the orientations of the areas on which they act, provided that the stress components referred to an arbitrary set of axes X, Y, Z are known at all these points.

Expanding the determinant (II.15) we arrive at the equation

$$\sigma^3 - c_2\sigma^2 + c_1\sigma - c_0 = 0, \tag{II.16}$$

whose coefficients

$$\begin{aligned} c_2 &= \sigma_{\xi\xi} + \sigma_{\eta\eta} + \sigma_{\zeta\zeta} = \sigma_1 + \sigma_2 + \sigma_3, \\ c_1 &= \sigma_{\xi\xi}\sigma_{\eta\eta} + \sigma_{\xi\xi}\sigma_{\zeta\zeta} + \sigma_{\eta\eta}\sigma_{\zeta\zeta} - \sigma_{\xi\eta}^2 \\ &\quad - \sigma_{\xi\zeta}^2 - \sigma_{\eta\zeta}^2 = \sigma_1\sigma_2 + \sigma_1\sigma_3 + \sigma_2\sigma_3, \\ c_0 &= \sigma_{\xi\xi}\sigma_{\eta\eta}\sigma_{\zeta\zeta} + 2\sigma_{\xi\eta}\sigma_{\xi\zeta}\sigma_{\eta\zeta} - \sigma_{\xi\xi}\sigma_{\eta\zeta}^2 - \sigma_{\eta\eta}\sigma_{\xi\zeta}^2 - \sigma_{\zeta\zeta}\sigma_{\xi\eta}^2 \\ &= \sigma_1\sigma_2\sigma_3 \end{aligned} \tag{II.17}$$

are invariant under change of axes, and represent invariants of the state of stress at every point of the deformed body.

§ 19. Conditions for Equilibrium of an Elementary Parallelopiped Isolated From a Deformed Body

From a deformed body, let us isolate an elementary rectangular parallelopiped whose edges are parallel to the X-, Y-, Z-axes and equal $d\xi$, $d\eta$, $d\zeta$, respectively (Fig. 13).

The faces of the element are characterized in Table 8.

Table 8

Name of Face	Coordinates of Centroid of its Area	Area of Face	Orientation of Face
AA_1B_1B	$\xi,\ \eta+\frac{d\eta}{2},\ \zeta+\frac{d\zeta}{2}$	$d\eta d\zeta$	$-\mathbf{i}_x$
DD_1C_1C	$\xi+d\xi,\eta+\frac{d\eta}{2},\ \zeta+\frac{d\zeta}{2}$	$d\eta d\zeta$	$+\mathbf{i}_x$
DD_1A_1A	$\xi+\frac{d\xi}{2},\eta,\zeta+\frac{d\zeta}{2}$	$d\xi d\zeta$	$-\mathbf{i}_y$
BB_1C_1C	$\xi+\frac{d\xi}{2},\eta+d\eta,\zeta+\frac{d\zeta}{2}$	$d\xi d\zeta$	$+\mathbf{i}_y$
$ADCB$	$\xi+\frac{d\xi}{2},\eta+\frac{d\eta}{2},\zeta$	$d\xi d\eta$	$-\mathbf{i}_z$
$A_1D_1C_1B_1$	$\xi+\frac{d\xi}{2},\eta+\frac{d\eta}{2},\zeta+d\zeta$	$d\xi d\eta$	$+\mathbf{i}_z$

Accordingly, the following surface forces (determined to within magnitudes of the third order) act on them:

$$AA_1B_1B \sim \sigma_{-\xi}(\xi,\ \eta+\tfrac{1}{2}d\eta,\ \zeta+\tfrac{1}{2}d\zeta)\,d\eta d\zeta$$

$$=-[\sigma_\xi(\xi,\eta,\zeta)+\frac{1}{2}\frac{\partial\sigma_\xi}{\partial\eta}d\eta+\frac{1}{2}\frac{\partial\sigma_\xi}{\partial\zeta}d\zeta]d\eta\, d\zeta,$$

$$DD_1C_1C \sim \sigma_\xi\left(\xi + d\xi,\ \eta + \frac{1}{2}d\eta,\ \zeta + \frac{1}{2}d\zeta\right)d\eta d\zeta$$
$$= \left[\sigma_\xi(\xi, \eta, \zeta) + \frac{\partial\sigma_\xi}{\partial\xi}d\xi + \frac{1}{2}\frac{\partial\sigma_\xi}{\partial\eta}d\eta + \frac{1}{2}\frac{\partial\sigma_\xi}{\partial\zeta}d\zeta\right]d\eta d\zeta,$$
$$DD_1A_1A \sim \sigma_{-\eta}\left(\xi + \frac{1}{2}d\xi,\ \eta,\ \zeta + \frac{1}{2}d\zeta\right)d\xi d\zeta$$
$$= -\left[\sigma_\eta(\xi, \eta, \zeta) + \frac{1}{2}\frac{\partial\sigma_\eta}{\partial\xi}d\xi + \frac{1}{2}\frac{\partial\sigma_\eta}{\partial\zeta}d\zeta\right]d\xi d\zeta,$$
$$BB_1C_1C \sim \sigma_\eta\left(\xi + \frac{1}{2}d\xi,\ \eta + d\eta,\ \zeta + \frac{1}{2}d\zeta\right)d\xi d\zeta$$
$$= \left[\sigma_\eta(\xi, \eta, \zeta) + \frac{1}{2}\frac{\partial\sigma_\eta}{\partial\xi}d\xi + \frac{\partial\sigma_\eta}{\partial\eta}d\eta + \frac{1}{2}\frac{\partial\sigma_\eta}{\partial\zeta}d\zeta\right]d\xi d\zeta, \quad \text{(II. 18)}$$
$$ADCB \sim \sigma_{-\zeta}\left(\xi + \frac{1}{2}d\xi,\ \eta + \frac{1}{2}d\eta,\ \zeta\right)d\xi d\eta$$
$$= -\left[\sigma_\zeta(\xi, \eta, \zeta) + \frac{1}{2}\frac{\partial\sigma_\zeta}{\partial\xi}d\xi + \frac{1}{2}\frac{\partial\sigma_\eta}{\partial\eta}d\eta\right]d\xi d\eta,$$
$$A_1D_1C_1B_1 \sim \sigma_\zeta\left(\xi + \frac{1}{2}d\xi,\ \eta + \frac{1}{2}d\eta,\ \zeta + d\zeta\right)d\xi d\zeta$$
$$= \left[\sigma_\zeta(\xi, \eta, \zeta) + \frac{1}{2}\frac{\partial\sigma_\zeta}{\partial\xi}d\xi + \frac{1}{2}\frac{\partial\sigma_\zeta}{\partial\eta}d\eta + \frac{\partial\sigma_\zeta}{\partial\zeta}d\zeta\right]d\xi d\eta.$$

In addition to surface forces, there will also be

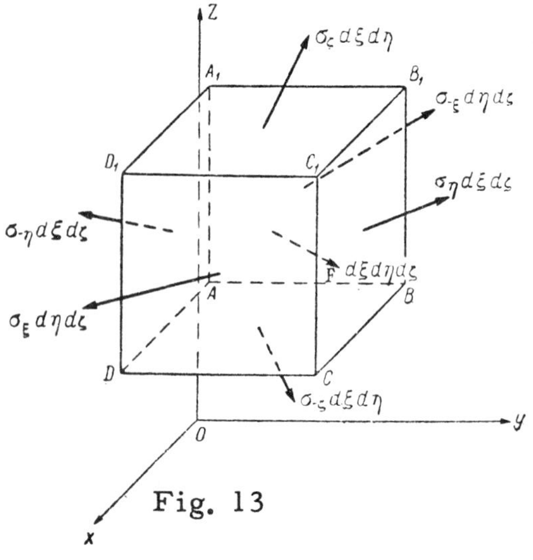

Fig. 13

body forces acting on the element. Their resultant (with the same accuracy as in (II.18)) is equal to

$$\mathbf{F}^*\left(\xi+\frac{1}{2}d\xi, \eta+\frac{1}{2}d\eta, \zeta+\frac{1}{2}d\zeta\right)d\xi d\eta d\zeta$$
$$=\left[\mathbf{F}(\xi, \eta, \zeta)+\frac{1}{2}\frac{\partial \mathbf{F}}{\partial \xi}d\xi+\frac{1}{2}\frac{\partial \mathbf{F}}{\partial \eta}d\eta+\frac{1}{2}\frac{\partial \mathbf{F}}{\partial \zeta}d\zeta\right]d\xi d\eta d\zeta$$
$$\approx \mathbf{F}^*(\xi, \eta, \zeta)\,d\xi d\eta d\zeta, \qquad \text{(II. 19)}$$

where $\mathbf{F}^*(\xi, \eta, \zeta)$ is the specific body force at the point $M^*(\xi, \eta, \zeta)$.

Summing all surface and body forces and setting the result equal to zero, we arrive at the following relation (after cancelling the common factor $d\xi d\eta d\zeta$):

$$\frac{\partial \sigma_\xi}{\partial \xi}+\frac{\partial \sigma_\eta}{\partial \eta}+\frac{\partial \sigma_\zeta}{\partial \zeta}+\mathbf{F}^*=0, \qquad \text{(II. 20)}$$

which is the condition resulting from setting equal to zero the resultant of all the forces acting on the element of volume isolated from the deformed body. The vector equation (II.20) is equivalent to the three scalar equations

$$\begin{aligned}
&\frac{\partial \sigma_{\xi\xi}}{\partial \xi}+\frac{\partial \sigma_{\eta\xi}}{\partial \eta}+\frac{\partial \sigma_{\zeta\xi}}{\partial \zeta}+F_\xi^*=0,\\
&\frac{\partial \sigma_{\xi\eta}}{\partial \xi}+\frac{\partial \sigma_{\eta\eta}}{\partial \eta}+\frac{\partial \sigma_{\zeta\eta}}{\partial \zeta}+F_\eta^*=0, \qquad \text{(II. 21)}\\
&\frac{\partial \sigma_{\xi\zeta}}{\partial \xi}+\frac{\partial \sigma_{\eta\zeta}}{\partial \eta}+\frac{\partial \sigma_{\zeta\zeta}}{\partial \zeta}+F_\zeta^*=0.
\end{aligned}$$

To obtain a complete solution of the problem of the equilibrium of the parallelopiped, it is necessary to supplement equations (II.21) by the conditions obtained by equating to zero the resultant of the moments of all the forces acting on the parallelopiped. The moment of these forces about the straight line parallel to the X-axis and passing through the center of the parallelopiped, is

$$M_\xi=\Sigma\,(y_i Z_i - z_i Y_i), \qquad \text{(II. 22)}$$

where Z_i and Y_i are the projections of an arbitrary force on the axes Z and Y, and y_i and z_i are the moment arms of these projections with respect to the above straight line. Applying this formula to all the forces (II.18) and (II.19), and retaining only magnitudes up to and including the third order, we obtain

$$M_{\xi} = (\sigma_{\eta\zeta} - \sigma_{\zeta\eta})\, d\xi d\eta d\zeta. \qquad \text{(II. 23)}$$

If we set this moment equal to zero, we get the equality

$$\sigma_{\eta\zeta} = \sigma_{\zeta\eta}. \qquad \text{(II. 24)}$$

Similarly, starting with the moments about the straight lines parallel to the Y- and Z-axes, it is possible to obtain the relations

$$\sigma_{\xi\eta} = \sigma_{\eta\xi}, \quad \sigma_{\xi\zeta} = \sigma_{\zeta\xi}. \qquad \text{(II. 25)}$$

Thus the condition obtained by equating to zero the resultant of the moments of the forces acting on an elementary parallelopiped reduces to the condition that the shearing stresses admit of a transposition of subscripts. This condition must be satisfied independently of the choice of directions of the X-, Y-, Z-axes; i. e., the stress components in any system of Cartesian coordinates must possess this property. We anticipated this result in § 18, where we made use of this property of the stress components.

Results (II.21) and (II.25) do not differ in form from the conditions for the equilibrium of a parallelopiped derived in the classical linear theory of elasticity. They differ in essence, however, because in the linear theory of elasticity, in deriving formulas (II.21) and (II.25), ξ, η, ζ are identified with x, y, and z, so that, when the forces are projected, the changes in position of the points of the body due to its deformation are neglected.

Thus, in the linear theory, no distinction is made between the predeformation and postdeformation values of the magnitudes and positions of the areas on which the stresses act; in other words, in projecting the forces, one neglects the rotation which an element of volume experiences as a result of the deformation. This, however, is far from admissible in all problems, and, in the general case, it is necessary to take into account the fact that the differentiation in (II.21) is performed not with respect to x, y, z (the coordinates of the points before deformation) but with respect to ξ, η, ζ (the coordinates of the points after deformation).

As we have noted, equations (II.21) are not more complicated superficially than the corresponding equations of the linear theory; in fact, however, they are considerably more complicated. The point is that the determination of the stresses and displacements under given loading and boundary conditions enters into the problem of the theory of elasticity. Here (according to (I.1)) the variables ξ, η, ζ, however, depend on u, v, w, and consequently the differentiation in (II.21) is performed with respect to parameters which depend on the functions to be determined, whereas in the linear theory of elasticity, the differentiation in the equations of equilibrium of an element is carried out with respect to parameters which do not involve unknown quantities. In this connection it is expedient to transform equations (II.21) in such a manner as to make explicit all the unknown functions implicit in them. In other words, it is desirable in these equations to pass from differentiation with respect to ξ, η, ζ to differentiation with respect to x, y, z. This will be accomplished in § 20.

§ 20. Transformation of the Equations of Equilibrium of an Element of Volume to the Cartesian Coordinates of the Points of the Body Before its Deformation

If we pass from differentiation with respect to ξ, η, ζ to differentiation with respect to x, y, z in (II.20), we get

$$\frac{\partial\sigma_\xi}{\partial x}\frac{\partial x}{\partial\xi}+\frac{\partial\sigma_\xi}{\partial y}\frac{\partial y}{\partial\xi}+\frac{\partial\sigma_\xi}{\partial z}\frac{\partial z}{\partial\xi}+\frac{\partial\sigma_\eta}{\partial x}\frac{\partial x}{\partial\eta}+\frac{\partial\sigma_\eta}{\partial y}\frac{\partial y}{\partial\eta}$$
$$+\frac{\partial\sigma_\eta}{\partial z}\frac{\partial z}{\partial\eta}+\frac{\partial\sigma_\zeta}{\partial x}\frac{\partial x}{\partial\zeta}+\frac{\partial\sigma_\zeta}{\partial y}\frac{\partial y}{\partial\zeta}+\frac{\partial\sigma_\zeta}{\partial z}\frac{\partial z}{\partial\zeta}+F^*=0. \quad (II.26)$$

The partial derivatives $\frac{\partial x}{\partial\xi}, \ldots, \frac{\partial z}{\partial\zeta}$ appearing here are determined by (I.14), according to which

$$\frac{\partial x}{\partial\xi}=\frac{\alpha_{11}}{D}, \quad \frac{\partial x}{\partial\eta}=\frac{\alpha_{12}}{D}, \quad \frac{\partial x}{\partial\zeta}=\frac{\alpha_{13}}{D},$$
$$\frac{\partial y}{\partial\xi}=\frac{\alpha_{21}}{D}, \quad \frac{\partial y}{\partial\eta}=\frac{\alpha_{22}}{D}, \quad \frac{\partial y}{\partial\zeta}=\frac{\alpha_{23}}{D}, \quad (II.27)$$
$$\frac{\partial z}{\partial\xi}=\frac{\alpha_{31}}{D}, \quad \frac{\partial z}{d\eta}=\frac{\alpha_{32}}{D}, \quad \frac{\partial z}{\partial\zeta}=\frac{\alpha_{33}}{D}.$$

Substituting these expressions in (II.26), we find that

$$\alpha_{11}\frac{\partial\sigma_\xi}{\partial x}+\alpha_{21}\frac{\partial\sigma_\xi}{\partial y}+\alpha_{31}\frac{\partial\sigma_\xi}{\partial z}+\alpha_{12}\frac{\partial\sigma_\eta}{\partial x}+\alpha_{22}\frac{\partial\sigma_\eta}{\partial y}$$
$$+\alpha_3\frac{\partial\sigma_\eta}{\partial z}+\alpha_{13}\frac{\partial\sigma_\zeta}{\partial x}+\alpha_{23}\frac{\partial\sigma_\zeta}{\partial y}+\alpha_{33}\frac{\partial\sigma_\zeta}{\partial z}+DF^*=0. \quad (II.28)$$

This equation may also be written in the form

$$\frac{\partial(\alpha_{11}\sigma_\xi+\alpha_{12}\sigma_\eta+\alpha_{13}\sigma_\zeta)}{\partial x}+\frac{\partial(\alpha_{21}\sigma_\xi+\alpha_{22}\sigma_\eta+\alpha_{23}\sigma_\zeta)}{\partial y}$$
$$+\frac{\partial(\alpha_{31}\sigma_\xi+\alpha_{32}\sigma_\eta+\alpha_{33}\sigma_\zeta)}{\partial z}-\left[\frac{\partial\alpha_{11}}{\partial x}+\frac{\partial\alpha_{21}}{\partial y}+\frac{\partial\alpha_{31}}{\partial z}\right]\sigma_\xi \qquad \text{(II.29)}$$
$$-\left[\frac{\partial\alpha_{12}}{\partial x}+\frac{\partial\alpha_{22}}{\partial y}+\frac{\partial\alpha_{32}}{\partial z}\right]\sigma_\eta$$
$$-\left[\frac{\partial\alpha_{13}}{\partial x}+\frac{\partial\alpha_{23}}{\partial y}+\frac{\partial\alpha_{33}}{\partial z}\right]\sigma_\zeta+DF^*=0.$$

If we replace the α_{ij} by their expressions in terms of u, v, w (in accordance with (I.15)), it is easy to see that the expressions in brackets in (II.29) are identically zero. Hence,

$$\frac{\partial(\alpha_{11}\sigma_\xi+\alpha_{12}\sigma_\eta+\alpha_{13}\sigma_\zeta)}{\partial x}+\frac{\partial(\alpha_{21}\sigma_\xi+\alpha_{22}\sigma_\eta+\alpha_{23}\sigma_\zeta)}{\partial y}$$
$$+\frac{\partial(\alpha_{31}\sigma_\xi+\alpha_{32}\sigma_\eta+\alpha_{33}\sigma_\zeta)}{\partial z}+DF^*=0. \qquad \text{(II.30)}$$

The combinations of stresses within the parentheses have a definite physical meaning which can be elucidated by the following considerations.

Let us suppose that a rectangular area perpendicular to the Z-axis and with sides dx, dy is isolated from the body before the deformation. As a result of the deformation, this area becomes a parallelogram, the directions of whose sides are given by the unit vectors $\mathbf{i}_1$, $\mathbf{i}_2$ (§ 2). Consequently the unit vector in the direction of the normal to the given area can be found from the equation

$$\mathbf{n}_3\sin(\mathbf{i}_1,\ \mathbf{i}_2)=\mathbf{i}_1\times\mathbf{i}_2. \qquad \text{(II.31)}$$

Projecting this on the X-, Y-, Z-axes, we obtain

$$\cos(X,\ \mathbf{n}_3)=\frac{1}{\sin(\mathbf{i}_1,\ \mathbf{i}_2)}\Big[\cos(Y,\ \mathbf{i}_1)\cos(Z,\ \mathbf{i}_2)$$
$$-\cos(Z,\ \mathbf{i}_1)\cos(Y,\ \mathbf{i}_2)\Big],$$
$$\cos(Y,\ \mathbf{n}_3)=\frac{1}{\sin(\mathbf{i}_1,\ \mathbf{i}_2)}\Big[\cos(Z,\ \mathbf{i}_1)\cos(X,\ i_2)$$
$$-\cos(X,\ \mathbf{i}_1)\cos(Z,\ \mathbf{i}_2)\Big], \qquad \text{(II.32)}$$

$$\cos(Z, \mathbf{n}_3) = \frac{1}{\sin(\mathbf{i}_1, \mathbf{i}_2)}\Big[\cos(X, \mathbf{i}_1)\cos(Y, \mathbf{i}_2) - \cos(Y, \mathbf{i}_1)\cos(X, \mathbf{i}_2)\Big],$$

but, according to (I.32) and (I.29),

$$\sin(\mathbf{i}_1, \mathbf{i}_2) = \cos\varphi_{xy} = \sqrt{1 - \frac{\varepsilon_{xy}^2}{(1+E_x)^2(1+E_y)^2}} = \frac{1}{(1+E_x)(1+E_y)}\sqrt{(1+2\varepsilon_{xx})(1+2\varepsilon_{yy}) - \varepsilon_{xy}^2}, \quad \text{(II.33)}$$

whereas the cosines entering into the brackets in (II.32) are given in Table 1 (p. 7).

We thus obtain

$$\cos(X, \mathbf{n}_3) = \frac{1}{\sqrt{(1+2\varepsilon_{xx})(1+2\varepsilon_{yy}) - \varepsilon_{xy}^2}} \times \left[\left(\frac{1}{2}e_{xy} + \omega_z\right)\left(\frac{1}{2}e_{yz} + \omega_x\right) - (1+e_{yy})\left(\frac{1}{2}e_{xz} - \omega_y\right)\right]$$

$$= \frac{\alpha_{31}}{\sqrt{(1+2\varepsilon_{xx})(1+2\varepsilon_{yy}) - \varepsilon_{xy}^2}}, \quad \text{(II.34)}$$

$$\cos(Y, \mathbf{n}_3) = \frac{\alpha_{32}}{\sqrt{(1+2\varepsilon_{xx})(1+2\varepsilon_{yy}) - \varepsilon_{xy}^2}},$$

$$\cos(Z, \mathbf{n}_3) = \frac{\alpha_{33}}{\sqrt{(1+2\varepsilon_{xx})(1+2\varepsilon_{yy}) - \varepsilon_{xy}^2}}.$$

Similarly one can show that

$$\cos(\mathbf{n}_1, X) = \frac{\alpha_{11}}{\sqrt{(1+2\varepsilon_{zz})(1+2\varepsilon_{yy}) - \varepsilon_{yz}^2}},$$

$$\cos(\mathbf{n}_1, Y) = \frac{\alpha_{12}}{\sqrt{(1+2\varepsilon_{zz})(1+2\varepsilon_{yy}) - \varepsilon_{yz}^2}},$$

$$\cos(\mathbf{n}_1, Z) = \frac{\alpha_{13}}{\sqrt{(1+2\varepsilon_{zz})(1+2\varepsilon_{yy}) - \varepsilon_{yz}^2}}, \quad \text{(II.35)}$$

$$\cos(X, \mathbf{n}_2) = \frac{\alpha_{21}}{\sqrt{(1+2\varepsilon_{xx})(1+2\varepsilon_{zz}) - \varepsilon_{xz}^2}},$$

$$\cos(Y, \mathbf{n}_2) = \frac{\alpha_{22}}{\sqrt{(1+2\varepsilon_{xx})(1+2\varepsilon_{zz}) - \varepsilon_{xz}^2}},$$

$$\cos(Z, \mathbf{n}_2) = \frac{\alpha_{23}}{\sqrt{(1+2\varepsilon_{xx})(1+2\varepsilon_{zz}) - \varepsilon_{xz}^2}},$$

where $\mathbf{n}_1$ and $\mathbf{n}_2$ are unit vectors in the directions of the normals to those areas of the deformed body which, before the deformation, were perpendicular to the X-axis and the Y-axis.

Note further that since the area $dxdy$, as a result of the deformation, becomes a parallelogram with sides $\mathbf{i}_1(1+E_x)\,dx$, $\mathbf{i}_2(1+E_y)\,dy$, the ratio of its area before deformation to its area after deformation is

$$\frac{S_z^*}{S_z} = (1+E_x)(1+E_y)\sin(\mathbf{i}_1, \mathbf{i}_2)$$
$$= \sqrt{(1+2\varepsilon_{xx})(1+2\varepsilon_{yy}) - \varepsilon_{xu}^2}. \qquad \text{(II.36)}$$

Analogously

$$\frac{S_y^*}{S_y} = \sqrt{(1+2\varepsilon_{xx})(1+2\varepsilon_{zz}) - \varepsilon_{xz}^2},$$
$$\frac{S_x^*}{S_x} = \sqrt{(1+2\varepsilon_{yy})(1+2\varepsilon_{zz}) - \varepsilon_{yz}^2}. \qquad \text{(II.37)}$$

Thus, the square roots entering into the denominators in (II.34), (II.35) are equal to the ratios of the areas of the elements of area of the body which are perpendicular to the X-, Y-, Z-axes before the deformation, to their areas after the deformation.

The determinant D, according to (I.94), represents the ratio of the volume of an element of volume of the body after the deformation, to its volume before the deformation:

$$D = \frac{V^*}{V}. \qquad \text{(II.38)}$$

If we take into account all the results obtained, as well as (II.6), which expresses the stress on an arbitrary area in terms of the stresses on areas perpendicular to the coordinate axes, we can rewrite (II.30) in the following final form:

$$\frac{\partial}{\partial x}\left(\frac{S_x^*}{S_x}\sigma_{n_1}\right)+\frac{\partial}{\partial y}\left(\frac{S_y^*}{S_y}\sigma_{n_2}\right)$$
$$+\frac{\partial}{\partial z}\left(\frac{S_z^*}{S_z}\sigma_{n_3}\right)+\frac{V^*}{V}\mathbf{F}^*=0. \qquad \text{(II.39)}$$

This form is assumed by (II.20) if the positions of the points of the deformed body are determined, not by the Cartesian coordinates ξ, η, ζ, but by the curvilinear coordinates $\tilde{x}, \tilde{y}, \tilde{z}$ (which are the Cartesian coordinates for the body in its initial state).

In the general case, the magnitude and direction of the specific body force may depend on the displacements u, v, w, so that

$$\mathbf{F}^*=\mathbf{F}^*(x, y, z, u, v, w). \qquad \text{(II.40)}$$

For example, if the body forces have an inertial character arising, say, from the rotation of the body about some axis, then clearly their magnitudes may change appreciably as a result of the deformation.

Resolve the vectors σ_{n_1}, σ_{n_2}, σ_{n_3} in the directions $\mathbf{i}_1$, $\mathbf{i}_2$, $\mathbf{i}_3$ and label their components as follows:

$$\begin{matrix} \sigma_{xx}, & \sigma_{xy}, & \sigma_{xz} \\ \sigma_{yx}, & \sigma_{yy}, & \sigma_{yz} \\ \sigma_{zx}, & \sigma_{zy}, & \sigma_{zz}. \end{matrix} \qquad \text{(II.41)}$$

Then (II.39) can be written

$$\frac{\partial}{\partial x}\left[\frac{S_x^*}{S_x}(\sigma_{xx}\mathbf{i}_1+\sigma_{xy}\mathbf{i}_2+\sigma_{xz}\mathbf{i}_3)\right]$$
$$+\frac{\partial}{\partial y}\left[\frac{S_y^*}{S_y}(\sigma_{yx}\mathbf{i}_1+\sigma_{yy}\mathbf{i}_2+\sigma_{yz}\mathbf{i}_3)\right]$$
$$+\frac{\partial}{\partial z}\left[\frac{S_z^*}{S_z}(\sigma_{zx}\mathbf{i}_1+\sigma_{zy}\mathbf{i}_2+\sigma_{zz}\mathbf{i}_3)\right]+\frac{V^*}{V}\mathbf{F}^*=0. \qquad \text{(II. 42)}$$

Projecting this vector relation on the X-, Y-, Z-axes and noting that the projections of the trihedral $\mathbf{i}_1$, $\mathbf{i}_2$, $\mathbf{i}_3$ on these axes are given in Table 1 in the preceding chapter, we have

$$\frac{\partial}{\partial x}\left[(1+e_{xx})\sigma^*_{xx}+\left(\frac{1}{2}e_{xy}-\omega_z\right)\sigma^*_{xy}+\left(\frac{1}{2}e_{xz}+\omega_y\right)\sigma^*_{xz}\right]$$
$$+\frac{\partial}{\partial y}\left[(1+e_{xx})\sigma^*_{yx}+\left(\frac{1}{2}e_{xy}-\omega_z\right)\sigma^*_{yy}+\left(\frac{1}{2}e_{xz}+\omega_y\right)\sigma^*_{yz}\right]$$
$$+\frac{\partial}{\partial z}\left[(1+e_{xx})\sigma^*_{zx}+\left(\frac{1}{2}e_{xy}-\omega_z\right)\sigma^*_{zy}+\left(\frac{1}{2}e_{xz}+\omega_y\right)\sigma^*_{zz}\right]$$
$$+\frac{V^*}{V}F_\xi^*=0, \tag{II.43}$$

$$\frac{\partial}{\partial x}\left[\left(\frac{1}{2}e_{xy}+\omega_z\right)\sigma^*_{xx}+(1+e_{yy})\sigma^*_{xy}+\left(\frac{1}{2}e_{yz}-\omega_x\right)\sigma^*_{xz}\right]$$
$$+\frac{\partial}{\partial y}\left[\left(\frac{1}{2}e_{xy}+\omega_z\right)\sigma^*_{yx}+(1+e_{yy})\sigma^*_{yy}+\left(\frac{1}{2}e_{yz}-\omega_x\right)\sigma^*_{yz}\right]$$
$$+\frac{\partial}{\partial z}\left[\left(\frac{1}{2}e_{xy}+\omega_z\right)\sigma^*_{zx}+(1+e_{yy})\sigma^*_{zy}+\left(\frac{1}{2}e_{yz}-\omega_x\right)\sigma^*_{zz}\right]$$
$$+\frac{V^*}{V}F^*_\eta=0,$$

$$\frac{\partial}{\partial x}\left[\left(\frac{1}{2}e_{xz}-\omega_y\right)\sigma^*_{xx}+\left(\frac{1}{2}e_{yz}+\omega_x\right)\sigma^*_{xy}+(1+e_{zz})\sigma^*_{xz}\right]$$
$$+\frac{\partial}{\partial y}\left[\left(\frac{1}{2}e_{xz}-\omega_y\right)\sigma^*_{yx}+\left(\frac{1}{2}e_{yz}+\omega_x\right)\sigma^*_{yy}+(1+e_{zz})\sigma^*_{yz}\right]$$
$$+\frac{\partial}{\partial z}\left[\left(\frac{1}{2}e_{xz}-\omega_y\right)\sigma^*_{zx}+\left(\frac{1}{2}e_{yz}+\omega_x\right)\sigma^*_{zy}+(1+e_{zz})\sigma^*_{zz}\right]$$
$$+\frac{V^*}{V}F_\zeta^*=0,$$

where

$$\sigma^*_{xx}=\frac{S^*_x}{S_x}\frac{\sigma_{xx}}{1+E_x},\quad \sigma^*_{xy}=\frac{S^*_x}{S_x}\frac{\sigma_{xy}}{1+E_y},\quad \sigma^*_{xz}=\frac{S^*_x}{S_x}\frac{\sigma_{xz}}{1+E_z},$$
$$\sigma^*_{yy}=\frac{S^*_y}{S_y}\frac{\sigma_{yy}}{1+E_y},\quad \sigma^*_{yz}=\frac{S^*_y}{S_y}\frac{\sigma_{yz}}{1+E_z},\quad \sigma^*_{yx}=\frac{S^*_y}{S_y}\frac{\sigma_{yx}}{1+E_x}, \tag{II.44}$$
$$\sigma^*_{zx}=\frac{S^*_z}{S_z}\frac{\sigma_{zx}}{1+E_x},\quad \sigma^*_{zy}=\frac{S^*_z}{S_z}\frac{\sigma_{zy}}{1+E_y},\quad \sigma^*_{zz}=\frac{S^*_z}{S_z}\frac{\sigma_{zz}}{1+E_z}.$$

The three equations (II.43) are equivalent to the vector equation (II.39), and constitute the scalar form of this equation.

The magnitudes σ^*_{ij} defined by (II.44) are not,

strictly speaking, stresses. They can be called stresses referred to the dimensions of an element of volume before, not after, the deformation.

If the deformations are small, then the ratios

$$\frac{S_j^*}{S_j}\frac{1}{1+E_i}\approx 1 \qquad \text{(II. 45)}$$

and σ_{ij}^* can be identified with σ_{ij}, i. e., with the stress components in the directions i_1, i_2, i_3. Thus, the difference between σ_{ij}^* and σ_{ij} becomes significant only under large deformations, when the shears and elongations cannot be neglected relative to unity.

We shall show in the next chapter that

$$\sigma_{ij}^* = \sigma_{ji}^*, \qquad \text{(II.46)}$$

i. e., that the subscripts in σ^* can be transposed. But then in the general case (under large deformations) the stresses σ_{ij} do not admit of such a transposition, as is clear from (II.44). This does not contradict the symmetry property of the stress matrix, derived in the preceding section. For, the property was established there only for stresses on three mutually perpendicular areas, resolved in the directions of the normals to these areas.

§ 21. Simplification of the Equations of Equilibrium in the Case of Small Elongations and Shears

The ratios $\frac{S_x^*}{S_x}$, $\frac{S_y^*}{S_y}$, $\frac{S_z^*}{S_z}$, $\frac{V^*}{V}$ differ from unity only by magnitudes of the same order as the elongations and shears. Hence, they can be set equal to unity for small deformations, and then (II.39) assumes the form

$$\frac{\partial \sigma_{n_1}}{\partial x}+\frac{\partial \sigma_{n_2}}{\partial y}+\frac{\partial \sigma_{n_3}}{\partial z}+\mathbf{F}^*=0. \qquad \text{(II.47)}$$

If we also carry out this simplification in the scalar equation (II.43), and, in addition, neglect the relative elongations E_x, E_y, E_z in comparison with unity, we obtain

$$\begin{aligned}
&\frac{\partial}{\partial x}\left[(1+e_{xx})\sigma_{xx}+\left(\frac{1}{2}e_{xy}-\omega_z\right)\sigma_{xy}+\left(\frac{1}{2}e_{xz}+\omega_y\right)\sigma_{xz}\right]\\
&+\frac{\partial}{\partial y}\left[(1+e_{xx})\sigma_{yx}+\left(\frac{1}{2}e_{xy}-\omega_z\right)\sigma_{yy}+\left(\frac{1}{2}e_{xz}+\omega_y\right)\sigma_{yz}\right]\\
&+\frac{\partial}{\partial z}\left[(1+e_{xx})\sigma_{zx}+\left(\frac{1}{2}e_{xy}-\omega_z\right)\sigma_{zy}\right.\\
&\qquad\left.+\left(\frac{1}{2}e_{xz}+\omega_y\right)\sigma_{zz}\right]+F^*_\xi=0,\\
&\frac{\partial}{\partial x}\left[\left(\frac{1}{2}e_{xy}+\omega_z\right)\sigma_{xx}+(1+e_{yy})\sigma_{xy}+\left(\frac{1}{2}e_{yz}-\omega_x\right)\sigma_{xz}\right]\\
&+\frac{\partial}{\partial y}\left[\left(\frac{1}{2}e_{xy}+\omega_z\right)\sigma_{yx}+(1+e_{yy})\sigma_{yy}+\left(\frac{1}{2}e_{yz}-\omega_x\right)\sigma_{yz}\right]\\
&+\frac{\partial}{\partial z}\left[\left(\frac{1}{2}e_{xy}+\omega_z\right)\sigma_{zx}+(1+e_{yy})\sigma_{zy}\right.\\
&\qquad\left.+\left(\frac{1}{2}e_{yz}-\omega_x\right)\sigma_{zz}\right]+F^*_\eta=0, \qquad \text{(II.48)}\\
&\frac{\partial}{\partial x}\left[\left(\frac{1}{2}e_{xz}-\omega_y\right)\sigma_{xx}+\left(\frac{1}{2}e_{yz}+\omega_x\right)\sigma_{xy}+(1+e_{zz})\sigma_{xz}\right]\\
&+\frac{\partial}{\partial y}\left[\left(\frac{1}{2}e_{xz}-\omega_y\right)\sigma_{yx}+\left(\frac{1}{2}e_{yz}+\omega_x\right)\sigma_{yy}+(1+e_{zz})\sigma_{yz}\right]\\
&+\frac{\partial}{\partial z}\left[\left(\frac{1}{2}e_{xz}-\omega_y\right)\sigma_{zx}+\left(\frac{1}{2}e_{yz}+\omega_x\right)\sigma_{zy}\right.\\
&\qquad\left.+(1+e_{zz})\sigma_{zz}\right]+F^*_\zeta=0.
\end{aligned}$$

Let us use a diagram to clarify the geometrical nature of the simplifications used to obtain these equations. Isolate a rectangular parallelopiped, with edges dx, dy, dz parallel to the X-, Y-, Z-axes, from the body before its deformation (Fig. 14). As a result of the deformation, this rectangular parallelopiped becomes an oblique one, with edges $(1+E_x)\,dx$, $(1+E_y)\,dy$, $(1+E_z)\,dz$ forming the angles $\frac{\pi}{2}-\varphi_{xy}$, $\frac{\pi}{2}-\varphi_{xz}$, $\frac{\pi}{2}-\varphi_{yz}$.

If, however, the angles of rotation are large relative to the shears φ_{xy}, φ_{xz}, φ_{yz}, then (as was already remarked in § 13 of Ch. I) the latter may be neglected in comparison with the former in projecting the forces. This means that the examined parallelopiped can also be represented by a rectangular one after the deformation (Fig. 14). Moreover, the smallness of the elongations and shears allows one (in examining the equilibrium of the parallelopiped) to ignore distinctions between its dimensions before and after the deformation, referring the stresses and body forces, respectively, to the initial dimensions of the areas and the initial volume of the element. It is thus permissible to represent the parallelopiped after the deformation, as equal to the parallelopiped before the deformation, but differing from it (geometrically) only in its position in space (Fig. 14).

On the basis of these remarks, the unit vectors $\mathbf{i}_1$, $\mathbf{i}_2$, $\mathbf{i}_3$ should be considered as mutually perpendicular (Fig. 14). Together they form (with the accuracy employed in the calculations) a trihedral of Cartesian axes, turned relative to the X-, Y-, Z-axes in accordance with the rotation experienced by a neighborhood of the examined point of the body as a result of the deformation.

The physical law which will be discussed in the next chapter establishes a relation between the strains and the stress components in the directions $\mathbf{i}_1$, $\mathbf{i}_2$, $\mathbf{i}_3$. In accordance with this, Fig. 14, in which the stress components along these directions are pictured, corresponds exactly to the system (II.48).

Summarizing, one can say that the equations

(II.48) were derived by assuming that in studying the equilibrium of an infinitesimal volume element of the body, one need only take the rotation of that element into account while its deformation may be neglected. We may therefore call the equations (II.48) the equilibrium condition for an

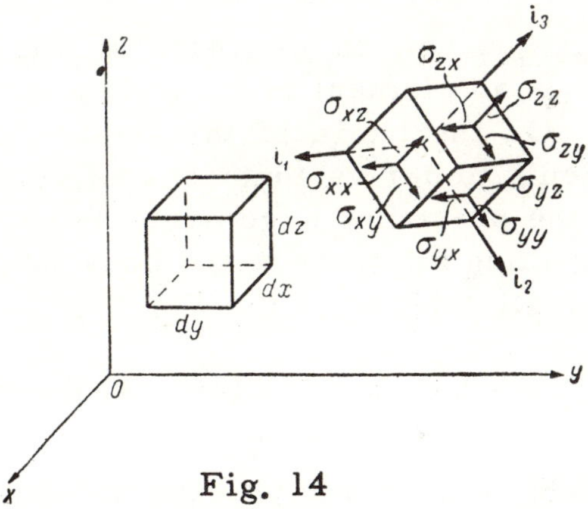

Fig. 14

infinitesimal volume element, valid under small relative deformations and arbitrary rotations.

§ 22. Simplification of the Equilibrium Equations for Small Rotations

If the angles of rotation are small compared to unity, then, by § 14, the parameters $e_{xx}, \ldots, e_{zz}$ differ from the strain components only by quantities of the same order as the squares of the angles of rotation. This enables us to simplify equations (II.48) by neglecting the strains and the squares of the angles of rotation as compared to the first powers of the latter. The result of this simplification is

$$\frac{\partial}{\partial x}[\sigma_{xx} - \omega_z\sigma_{xy} + \omega_y\sigma_{xz}] + \frac{\partial}{\partial y}[\sigma_{yx} - \omega_z\sigma_{yy} + \omega_y\sigma_{yz}] +$$

$$+\frac{\partial}{\partial z}[\sigma_{zx}-\omega_z\sigma_{zy}+\omega_y\sigma_{zz}]+\overset{*}{F}_\xi=0,$$

$$\frac{\partial}{\partial x}[\omega_z\sigma_{xx}+\sigma_{xy}-\omega_x\sigma_{xz}]+\frac{\partial}{\partial y}[\omega_z\sigma_{yx}+\sigma_{yy}-\omega_x\sigma_{yz}]$$
$$+\frac{\partial}{\partial z}[\omega_z\sigma_{zx}+\sigma_{zy}-\omega_x\sigma_{zz}]+\overset{*}{F}_\eta=0, \qquad \text{(II.49)}$$

$$\frac{\partial}{\partial x}[-\omega_y\sigma_{xx}+\omega_x\sigma_{xy}+\sigma_{xz}]+\frac{\partial}{\partial y}[-\omega_y\sigma_{yx}+\omega_x\sigma_{yy}+\sigma_{yz}]$$
$$+\frac{\partial}{\partial z}[-\omega_y\sigma_{zx}+\omega_x\sigma_{zy}+\sigma_{zz}]+\overset{*}{F}_\zeta=0,$$

where ω_x, ω_y, ω_z can be regarded as the angles of rotation of an element of the body about the axes X, Y, Z. On the basis of this last remark, it is not difficult to derive equations (II.49) independently of the preceding arguments by using Fig. 14 and projecting all surface and body forces acting on the parallelopiped, on the X-, Y-, Z-axes.

§ 23. Transition to the Classical Equations of Equilibrium

The next step in simplifying the equations of equilibrium of a volume element consists in assuming that the angles of rotation are so small that the terms in equations (II.49) which contain them as factors can be neglected in comparison with the terms which do not.

Equations (II.49) then reduce to

$$\frac{\partial\sigma_{xx}}{\partial x}+\frac{\partial\sigma_{yx}}{\partial y}+\frac{\partial\sigma_{zx}}{\partial z}+F_\xi=0,$$

$$\frac{\partial\sigma_{xy}}{\partial x}+\frac{\partial\sigma_{yy}}{\partial y}+\frac{\partial\sigma_{zy}}{\partial z}+F_\eta=0, \qquad \text{(II.50)}$$

$$\frac{\partial\sigma_{xz}}{\partial x}+\frac{\partial\sigma_{yz}}{\partial y}+\frac{\partial\sigma_{zz}}{\partial z}+F_\zeta=0.$$

It is clear that writing the equations of equilibrium in this form is equivalent to neglecting the rotations of the volume element when all the forces acting on it are projected, i. e., by identifying the directions $\mathbf{i}_1, \mathbf{i}_2, \mathbf{i}_3$ with X, Y, Z. In that case, however, the stress components $\sigma_{xx}, \ldots, \sigma_{yz}$ in the directions of the local trihedral of the curvilinear coordinate system $\mathbf{i}_1, \mathbf{i}_2, \mathbf{i}_3$ are identical with $\sigma_{\xi\xi}, \ldots, \sigma_{\eta\zeta}$, the stress components along the X-, Y-, Z-axes. Hence (II.50) can also be written in the form

$$\begin{aligned} &\frac{\partial \sigma_{\xi\xi}}{\partial x} + \frac{\partial \sigma_{\eta\xi}}{\partial y} + \frac{\partial \sigma_{\zeta\xi}}{\partial z} + F_\xi = 0, \\ &\frac{\partial \sigma_{\xi\eta}}{\partial x} + \frac{\partial \sigma_{\eta\eta}}{\partial y} + \frac{\partial \sigma_{\zeta\eta}}{\partial z} + F_\eta = 0, \\ &\frac{\partial \sigma_{\xi\zeta}}{\partial x} + \frac{\partial \sigma_{\eta\zeta}}{\partial y} + \frac{\partial \sigma_{\zeta\zeta}}{\partial z} + F_\zeta = 0. \end{aligned} \tag{II.51}$$

The three equations (II.51) can be replaced by the single vector equation

$$\frac{\partial \sigma_\xi}{\partial x} + \frac{\partial \sigma_\eta}{\partial y} + \frac{\partial \sigma_\zeta}{\partial z} + \mathbf{F} = 0, \tag{II.52}$$

which, combined with equations of the form

$$\sigma_{ij} = \sigma_{ji}$$

are the conditions of equilibrium for a volume element in the classical theory of elasticity. Equation (II.52) follows from (II.20) if, in the latter, differentiation with respect to ξ, η, ζ is replaced by differentiation with respect to x, y, z.

Equation (II.52) has been obtained by a succession of simplifications of the exact equation (II.39). It is seen from the nature of these simplifications that their validity rests on the smallness of the elongations and shears and on the smallness of the angles of rotation. It is essential to note that there are problems in which, even for rotations which are small compared to unity, it is not possible to pass from (II.49) to (II.50). In order to show this, let us examine, for example, the expressions

$$\begin{aligned} &\sigma_{xz} - \omega_y \sigma_{xx} + \omega_x \sigma_{xy}, \\ &\sigma_{yz} - \omega_y \sigma_{yx} + \omega_x \sigma_{yy}, \\ &\sigma_{zz} - \omega_y \sigma_{zx} + \omega_x \sigma_{zy}, \end{aligned} \qquad \text{(II.53)}$$

which enter into the third equation of the system (II.49). It is clear that whether or not the non-linear terms in these expressions may be neglected depends not only on the magnitude of the ω_j's but also on the comparative magnitudes of the σ_{ij}'s. Thus if the stresses σ_{xz}, σ_{yz}, σ_{zz} are considerably smaller than the stresses σ_{xx}, σ_{yx}, σ_{zx}, the linearization of the expressions (II.53) may be invalid, even though all the ω_j's are small in comparison with unity. It is therefore impossible to linearize the third equation of the system (II.49).

It follows that the smallness of the angles of rotation in comparison with unity is not a sufficient condition for the linearization of the equations of equilibrium. It is also essential to know whether the stresses which are multiplied by rotations are large in comparison with those stresses which enter linearly into the equations. The problems of elastic stability, of the bending of thin plates, and others, are cases in point.

§ 24. Transition to Curvilinear Coordinates

In the preceding discussion the points of the body were referred to a Cartesian coordinate system. Such a coordinate system is convenient for bodies which are bounded by mutually perpendicular planes, but is much less convenient if the body is bounded by curved surfaces. Indeed, in first case the equations of the boundaries of the body are very simple in Cartesian coordinates, while in the second case they are much more complicated. This makes the imposition of boundary conditions on the solution more difficult.

Hence it is more expedient in many cases to use curvilinear, in place of Cartesian, coordinates. The curvilinear coordinates should always be selected in such a way that the bounding surfaces of the body should at the same time be also coordinate surfaces. This will result in an especially simple formulation of the boundary conditions. In this connection we shall supplement the results already obtained in this chapter with a discussion of the conditions of equilibrium for a body whose points are referred to an arbitrary orthogonal curvilinear coordinate system $\alpha_1, \alpha_2, \alpha_3$.

To shorten the calculations involved in this transformation we may use the fact, already noted, that the equations of equilibrium of a volume element in the nonlinear theory are similar in appearance to the corresponding equations of the classical theory.

Indeed, in the nonlinear theory the conditions of equilibrium for an element referred to Cartesian coordinates reduce to the equation

$$\frac{\partial}{\partial x}\left(\frac{S_x^*}{S_x}\sigma_{n_1}\right)+\frac{\partial}{\partial y}\left(\frac{S_y^*}{S_y}\sigma_{n_2}\right)+\frac{\partial}{\partial z}\left(\frac{S_z^*}{S_z}\sigma_{n_3}\right)+\frac{V^*}{V}\mathbf{F}^*=0. \quad \text{(II.54)}$$

This, in the linear theory, assumes the form

$$\frac{\partial \sigma_x}{\partial x}+\frac{\partial \sigma_y}{\partial y}+\frac{\partial \sigma_z}{\partial z}+\mathbf{F}^*=0 \quad \text{(II.55)}$$

which is obtained by substituting in equation (II.54) for the vectors $\frac{S_x^*}{S_x}\sigma_{n_1}, \frac{S_y^*}{S_y}\sigma_{n_2}, \frac{S_z^*}{S_z}\sigma_{n_3}$ the vectors

$$\sigma_x, \sigma_y, \sigma_z.$$

Hence it is clear that if the transformation of equation (II.55) to an arbitrary orthogonal curvilinear coordinate system is known, then the corresponding transformation of equation (II.54) can be written down by analogy.

If the points of the body are referred to curvilinear coordinates (I, § 16), we isolate an infinitesimal volume element which is bounded by the six coordinate surfaces of the curvilinear system chosen. As a result of the deformation, this element changes its position in space (due to displacement and rotation) and, moreover, changes its dimensions and form. Its edges, initially equal to $\mathbf{k}_1 H_1 d\alpha_1$, $\mathbf{k}_2 H_2 d\alpha_2$, $\mathbf{k}_3 H_3 d\alpha_3$, now become

$$\mathbf{k}_1' H_1 (1 + E_{\alpha_1})\, d\alpha_1, \quad \mathbf{k}_2' H_2 (1 + E_{\alpha_2})\, d\alpha_2, \quad \mathbf{k}_3' H_3 (1 + E_{\alpha_3})\, d\alpha_3,$$

where $\mathbf{k}_1', \mathbf{k}_2', \mathbf{k}_3'$ are the unit vectors in the directions of the linear elements which, in the unstrained state, coincided with the vectors $\mathbf{k}_1$, $\mathbf{k}_2$, $\mathbf{k}_3$. The cosines of the angles between the trihedrals $\mathbf{k}_1, \mathbf{k}_2, \mathbf{k}_3$ and $\mathbf{k}_1', \mathbf{k}_2', \mathbf{k}_3'$ are given in Table 1 (Ch. I), in which the values of the parameters e and ω are determined from (I.122) and (I.124).

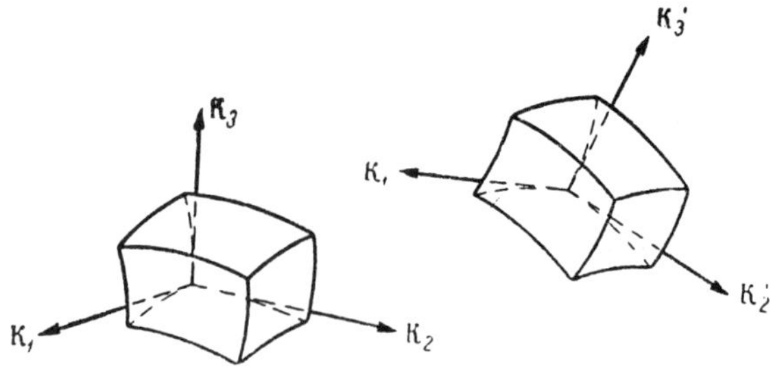

Fig. 15

In analogy with the resolution of the stresses acting on the faces of an element in the directions $\mathbf{i}_1$, $\mathbf{i}_2$, $\mathbf{i}_3$ in the Cartesian system, we now resolve the stresses acting on the new element in the directions $\mathbf{k}_1'$, $\mathbf{k}_2'$, $\mathbf{k}_3'$.

In books on the classical theory of elasticity

(see, e. g., Love's Mathematical Theory of Elasticity, p. 90) it is proved that, in an orthogonal curvilinear coordinate system, equation (II.55) is replaced by the following three scalar equations:

$$\begin{aligned}
&\frac{1}{H_1H_2H_3}\left\{\frac{\partial}{\partial\alpha_1}(H_2H_3\widehat{\alpha_1\alpha_1})+\frac{\partial}{\partial\alpha_2}(H_3H_1\widehat{\alpha_2\alpha_1})+\frac{\partial}{\partial\alpha_3}(H_1H_2\widehat{\alpha_3\alpha_1})\right\}\\
&\quad+\frac{1}{H_1H_2}\frac{\partial H_1}{\partial\alpha_2}\widehat{\alpha_1\alpha_2}+\frac{1}{H_1H_3}\frac{\partial H_1}{\partial\alpha_3}\widehat{\alpha_1\alpha_3}-\frac{1}{H_1H_2}\frac{\partial H_2}{\partial\alpha_1}\widehat{\alpha_2\alpha_2}\\
&\qquad-\frac{1}{H_1H_3}\frac{\partial H_3}{\partial\alpha_1}\widehat{\alpha_3\alpha_3}+F_{\alpha 1}=0
\end{aligned}$$

$$\begin{aligned}
&\frac{1}{H_1H_2H_3}\left\{\frac{\partial}{\partial\alpha_1}(H_2H_3\widehat{\alpha_1\alpha_2})+\frac{\partial}{\partial\alpha_2}(H_3H_1\widehat{\alpha_2\alpha_2})+\frac{\partial}{\partial\alpha_3}(H_1H_2\widehat{\alpha_3\alpha_2})\right\}\\
&\quad+\frac{1}{H_2H_3}\frac{\partial H_2}{\partial\alpha_3}\widehat{\alpha_2\alpha_3}+\frac{1}{H_2H_1}\frac{\partial H_2}{\partial\alpha_1}\widehat{\alpha_2\alpha_1}-\frac{1}{H_2H_3}\frac{\partial H_3}{\partial\alpha_2}\widehat{\alpha_3\alpha_3}\\
&\qquad-\frac{1}{H_2H_1}\frac{\partial H_1}{\partial\alpha_2}\widehat{\alpha_1\alpha_1}+F_{\alpha_2}=0,
\end{aligned}\tag{II.56}$$

$$\begin{aligned}
&\frac{1}{H_1H_2H_3}\left\{\frac{\partial}{\partial\alpha_1}(H_2H_3\widehat{\alpha_1\alpha_3})+\frac{\partial}{\partial\alpha_2}(H_3H_1\widehat{\alpha_2\alpha_3})+\frac{\partial}{\partial\alpha_3}(H_1H_2\widehat{\alpha_3\alpha_3})\right\}\\
&\quad+\frac{1}{H_3H_1}\frac{\partial H_3}{\partial\alpha_1}\widehat{\alpha_3\alpha_1}+\frac{1}{H_3H_2}\frac{\partial H_3}{\partial\alpha_2}\widehat{\alpha_3\alpha_2}-\frac{1}{H_3H_1}\frac{\partial H_1}{\partial\alpha_3}\widehat{\alpha_1\alpha_1}\\
&\qquad-\frac{1}{H_3H_2}\frac{\partial H_2}{\partial\alpha_3}\widehat{\alpha_2\alpha_2}+F_{\alpha_3}=0.
\end{aligned}$$

Here, H_1, H_2, H_3 are the Lame coefficients (Ch. I, § 16);

$F_{\alpha_1}, F_{\alpha_2}, F_{\alpha_3}$ are the projections of the specific body force on the directions α_1, α_2, α_3 ;

$\sigma_{\alpha_1}, \sigma_{\alpha_2}, \sigma_{\alpha_3}$ are the stresses on the areas perpendicular to the dihedrals $[\mathbf{k}_2', \mathbf{k}_3']$, $[\mathbf{k}_3', \mathbf{k}_1']$, $[\mathbf{k}_1', \mathbf{k}_2']$;

$\widehat{\alpha_1\alpha_1}, \widehat{\alpha_1\alpha_2}, \widehat{\alpha_1\alpha_3}$ are the components of the stress σ_{α_1} along $\mathbf{k}_1, \mathbf{k}_2, \mathbf{k}_3$;

$\widehat{\alpha_2\alpha_1}, \widehat{\alpha_2\alpha_2}, \widehat{\alpha_2\alpha_3}$ are the components of the stress σ_{α_2} along $\mathbf{k}_1, \mathbf{k}_2, \mathbf{k}_3$;

$\widehat{\alpha_3\alpha_1}, \widehat{\alpha_3\alpha_2}, \widehat{\alpha_3\alpha_3}$ are the components of the stress σ_{α_3} along $\mathbf{k}_1, \mathbf{k}_2, \mathbf{k}_3$.

In the linear theory no distinction is made between $\mathbf{k}_1, \mathbf{k}_2, \mathbf{k}_3$ and $\mathbf{k}_1', \mathbf{k}_2', \mathbf{k}_3'$.

Turning now to the nonlinear theory and taking into account the similarity of equations (II.54) and (II.55), we may conclude that the equations in the coordinates α_1, α_2, α_3 which correspond to (II.54) must be identical in form to the system (II.56). However, in these equations the expressions

$$\widehat{\alpha_1\alpha_1},\ \widehat{\alpha_1\alpha_2},\ \widehat{\alpha_1\alpha_3},$$

$$\widehat{\alpha_2\alpha_1},\ \widehat{\alpha_2\alpha_2},\ \widehat{\alpha_3\alpha_2},$$

$$\widehat{\alpha_3\alpha_1},\ \widehat{\alpha_2\alpha_3},\ \widehat{\alpha_3\alpha_3}$$

must be replaced by the components, along the directions of the axes of the local trihedral, $\mathbf{k}_1$, $\mathbf{k}_2$, $\mathbf{k}_3$, of the vectors $\frac{\overset{*}{S}_1}{S_1}\sigma_{n_1}$, $\frac{\overset{*}{S}_2}{S_2}\sigma_{n_2}$, $\frac{\overset{*}{S}_3}{S_3}\sigma_{n_3}$, where $\mathbf{n}_1=[\mathbf{k}_2', \mathbf{k}_3']$, $\mathbf{n}_2=[\mathbf{k}_3', \mathbf{k}_1']$, $\mathbf{n}_3=[\mathbf{k}_1', \mathbf{k}_2']$. Denoting the components of σ_{n_1}, σ_{n_2}, σ_{n_3} along $\mathbf{k}_1'$, $\mathbf{k}_2'$, $\mathbf{k}_3'$, by

σ_{11}, σ_{12}, σ_{13} for the components of σ_{n_1},
σ_{21}, σ_{22}, σ_{23} for the components of σ_{n_2},
σ_{31}, σ_{32}, σ_{33} for the components of σ_{n_3},

we obtain

$$\begin{aligned}
\frac{\overset{*}{S}_1}{S_1}\sigma_{n_1} &= \frac{\overset{*}{S}_1}{S_1}(\sigma_{11}\mathbf{k}_1'+\sigma_{12}\mathbf{k}_2'+\sigma_{13}\mathbf{k}_3'),\\
\frac{\overset{*}{S}_2}{S_2}\sigma_{n_2} &= \frac{\overset{*}{S}_2}{S_2}(\sigma_{21}\mathbf{k}_1'+\sigma_{22}\mathbf{k}_2'+\sigma_{23}\mathbf{k}_3'),\\
\frac{\overset{*}{S}_3}{S_3}\sigma_{n_3} &= \frac{\overset{*}{S}_3}{S_3}(\sigma_{31}\mathbf{k}_1'+\sigma_{32}\mathbf{k}_2'+\sigma_{33}\mathbf{k}_3').
\end{aligned} \tag{II.57}$$

Taking this into account, as well as the rule for projecting the vectors $\mathbf{k}_1'$, $\mathbf{k}_2'$, $\mathbf{k}_3'$ on $\mathbf{k}_1$, $\mathbf{k}_2$, $\mathbf{k}_3$ given in Table 1 (Ch. I), we conclude that the system of scalar equations in the orthogonal curvilinear coordinate system α_1, α_2, α_3, corresponding to the vector equation (II.54), can be obtained from (II.56) by replacing $\widehat{\alpha_1\alpha_1}, \ldots, \widehat{\alpha_3\alpha_3}$ by the following expressions:

$$\begin{aligned}
\widehat{\alpha_1\alpha_1} &= (1+e_{11})\overset{*}{\sigma}_{11} + \left(\frac{1}{2}e_{12} - \omega_3\right)\overset{*}{\sigma}_{12} + \left(\frac{1}{2}e_{13} + \omega_2\right)\overset{*}{\sigma}_{13},\\
\widehat{\alpha_1\alpha_2} &= \left(\frac{1}{2}e_{12} + \omega_3\right)\overset{*}{\sigma}_{11} + (1+e_{22})\overset{*}{\sigma}_{12} + \left(\frac{1}{2}e_{23} - \omega_1\right)\overset{*}{\sigma}_{13},\\
\widehat{\alpha_1\alpha_3} &= \left(\frac{1}{2}e_{13} - \omega_2\right)\overset{*}{\sigma}_{11} + \left(\frac{1}{2}e_{23} + \omega_1\right)\overset{*}{\sigma}_{12} + (1+e_{33})\overset{*}{\sigma}_{13},\\
\widehat{\alpha_2\alpha_1} &= (1+e_{11})\overset{*}{\sigma}_{21} + \left(\frac{1}{2}e_{12} - \omega_3\right)\overset{*}{\sigma}_{22} + \left(\frac{1}{2}e_{13} + \omega_2\right)\overset{*}{\sigma}_{23},\\
\widehat{\alpha_2\alpha_2} &= \left(\frac{1}{2}e_{12} + \omega_3\right)\overset{*}{\sigma}_{21} + (1+e_{22})\overset{*}{\sigma}_{22} + \left(\frac{1}{2}e_{23} - \omega_1\right)\overset{*}{\sigma}_{23},\\
\widehat{\alpha_2\alpha_3} &= \left(\frac{1}{2}e_{13} - \omega_2\right)\overset{*}{\sigma}_{21} + \left(\frac{1}{2}e_{23} + \omega_1\right)\overset{*}{\sigma}_{22} + (1+e_{33})\overset{*}{\sigma}_{23},
\end{aligned} \tag{II.58}$$

$$\begin{aligned}
\widehat{\alpha_3\alpha_1} &= (1+e_{11})\overset{*}{\sigma}_{31} + \left(\frac{1}{2}e_{12} - \omega_3\right)\overset{*}{\sigma}_{32} + \left(\frac{1}{2}e_{13} + \omega_2\right)\overset{*}{\sigma}_{33},\\
\widehat{\alpha_3\alpha_2} &= \left(\frac{1}{2}e_{12} + \omega_3\right)\overset{*}{\sigma}_{31} + (1+e_{22})\overset{*}{\sigma}_{32} + \left(\frac{1}{2}e_{23} - \omega_1\right)\overset{*}{\sigma}_{33},\\
\widehat{\alpha_3\alpha_3} &= \left(\frac{1}{2}e_{13} - \omega_2\right)\overset{*}{\sigma}_{31} + \left(\frac{1}{2}e_{23} + \omega_1\right)\overset{*}{\sigma}_{32} + (1+e_{33})\overset{*}{\sigma}_{33}.
\end{aligned}$$

Here $\overset{*}{\sigma}_{11}, \ldots, \overset{*}{\sigma}_{33}$ are related to $\sigma_{11}, \ldots, \sigma_{33}$ by equations of the form (II.44) with

$$\begin{aligned}
\frac{\overset{*}{S}_1}{S_1} &= \sqrt{(1+2\varepsilon_{22})(1+2\varepsilon_{33}) - \varepsilon_{23}^2},\\
\frac{\overset{*}{S}_2}{S_2} &= \sqrt{(1+2\varepsilon_{11})(1+2\varepsilon_{33}) - \varepsilon_{13}^2},\\
\frac{\overset{*}{S}_3}{S_3} &= \sqrt{(1+2\varepsilon_{11})(1+2\varepsilon_{22}) - \varepsilon_{12}^2},
\end{aligned} \tag{II.59}$$

while the parameters $e_{11}, \ldots, \omega_1, \ldots, \varepsilon_{11}, \ldots$ can be expressed in terms of the components of the displacements u, v, w in the directions $\mathbf{k}_1, \mathbf{k}_2, \mathbf{k}_3$ by equations (I.122), (I.124), (I.125).

To transform equations (II.56) into the equations of the nonlinear theory, besides replacing the stresses, it is also necessary to replace F_{α_1}, F_{α_2}, F_{α_3}, respectively, by

$$\frac{V^*}{V}F^*_{\alpha_1},\ \frac{V^*}{V}F^*_{\alpha_2},\ \frac{V^*}{V}F^*_{\alpha_3},\quad \text{where}\quad F^*_{\alpha_1},\ F^*_{\alpha_2},\ F^*_{\alpha_3}$$

are the projections on $\mathbf{k}_1$, $\mathbf{k}_2$, $\mathbf{k}_3$ of the specific body forces relative to the strained body; while

$$\frac{V^*}{V} = D = (1 + \Delta), \tag{II.60}$$

where Δ is the volume increment (Ch. I, § 11).

The above rules for transforming the system (II.43) to orthogonal curvilinear coordinates were established without neglecting any terms. Hence substitution of equations (II.58) into (II.56) will make the latter correspond precisely to (II.43).

If the elongations and shears are negligibly small compared to unity, equations (II.58) can be simplified by identifying $\overset{*}{\sigma}_{11}$, $\overset{*}{\sigma}_{12}$, ..., $\overset{*}{\sigma}_{33}$ with the stress components σ_{11}, σ_{12}, ..., σ_{33} (§ 20).

If, in addition, the angles of rotation are small compared to unity, these equations may be further simplified to yield

$$\begin{aligned}
\widehat{\alpha_1\alpha_1} &= \sigma_{11} - \omega_3\sigma_{12} + \omega_2\sigma_{13},\\
\widehat{\alpha_1\alpha_2} &= \sigma_{12} + \omega_3\sigma_{11} - \omega_1\sigma_{13},\\
\widehat{\alpha_1\alpha_3} &= \sigma_{13} + \omega_1\sigma_{12} - \omega_2\sigma_{11},\\
\widehat{\alpha_2\alpha_1} &= \sigma_{21} - \omega_3\sigma_{22} + \omega_2\sigma_{23},\\
\widehat{\alpha_2\alpha_2} &= \sigma_{22} + \omega_3\sigma_{21} - \omega_1\sigma_{23},\\
\widehat{\alpha_2\alpha_3} &= \sigma_{23} + \omega_1\sigma_{22} - \omega_2\sigma_{21},\\
\widehat{\alpha_3\alpha_1} &= \sigma_{31} - \omega_3\sigma_{32} + \omega_2\sigma_{33},\\
\widehat{\alpha_3\alpha_2} &= \sigma_{32} + \omega_3\sigma_{31} - \omega_1\sigma_{33},\\
\widehat{\alpha_3\alpha_3} &= \sigma_{33} + \omega_1\sigma_{32} - \omega_2\sigma_{31}.
\end{aligned} \tag{II.61}$$

Finally, if the angles of rotation are small quantities of the same order of magnitude as the strain components, the products of the stresses by the angles of rotation may be neglected in equations (II.61). This leads to the equations

$$\widehat{\alpha_1\alpha_1} = \sigma_{11}, \quad \widehat{\alpha_1\alpha_2} = \sigma_{12}, \quad \widehat{\alpha_1\alpha_3} = \sigma_{13},$$
$$\widehat{\alpha_2\alpha_1} = \sigma_{21}, \quad \widehat{\alpha_2\alpha_2} = \sigma_{22}, \quad \widehat{\alpha_2\alpha_3} = \sigma_{23} \qquad \text{(II.62)}$$
$$\widehat{\alpha_3\alpha_1} = \sigma_{31}, \quad \widehat{\alpha_3\alpha_2} = \sigma_{32}, \quad \widehat{\alpha_3\alpha_3} = \sigma_{33}.$$

In that case, equations (II.56) become identical with the equations of equilibrium of the linear theory referred to the orthogonal curvilinear coordinates $\alpha_1, \alpha_2, \alpha_3$.

CHAPTER III

STRAIN ENERGY, BOUNDARY CONDITIONS, STRESS-STRAIN LAW

§ 25. Strain Energy

The system of differential equations derived in the last chapter, which expresses the conditions of equilibrium for a volume element isolated from a solid body in a state of strain, contains more unknowns than equations. Indeed, it consists of six equations ((II.43) and (II.46)) containing twelve unknowns (nine stress and three displacement components).

Hence the problem of the equilibrium of a deformed solid body remains indeterminate until six supplementary equations are established. These relate the stress components to the displacement components and express the law according to which the material of the given body resists various forms of deformation. A theoretical explanation of this law would necessarily require an insight into the nature of intermolecular forces which seek to keep the particles of the solid body at definite distances from one another. The present state of scientific development, however, offers no adequate solution to this difficult problem. Hence, at the present time, the relation between strains and stresses, which differs for different materials, is established mainly by experiment. Some general properties inherent in this relation can, however, be explained theoretically. Let us begin by assuming that the process of deformation is isothermal and that the work expended on changing the volume and form of an arbitrary infinitesimal rectangular parallelopiped isolated from the body is independent of the manner in which the transition from the initial state of this element to the strained state is realized. We shall assume, in other words, that the role of the dissipative (nonconservative) forces in the process of interaction of the particles of the body undergoing deformation is negligible compared to the role of the conservative forces.

A body which satisfies this assumption must return to its initial dimensions and form after the load on it is removed. In this respect the body may be called ideally elastic. The work required to deform an elementary parallelopiped of an elastic body can be expressed in the form

$$dA = \mathfrak{I}(\varepsilon_{xx}, \varepsilon_{yy}, \varepsilon_{zz}, \varepsilon_{xy}, \varepsilon_{xz}, \varepsilon_{yz})\, dx dy dz, \qquad \text{(III.1)}$$

i. e., it is equal to the product of the initial volume of the parallelopiped and a certain function of the six strain components. The form of this function depends on the physical properties of the given material, but is independent of the dimensions and shape of the body. On the other hand, the strain components can always be expressed in terms of the three principal strain components $\varepsilon_1, \varepsilon_2, \varepsilon_3$ and the direction cosines of the principal axes of strain $\vec{\varepsilon}_1, \vec{\varepsilon}_2, \vec{\varepsilon}_3$ with respect to the X-, Y-, Z-axes (Ch.I, § 4,5). Here, the direction cosines can be regarded as functions of three independent quantities, e. g., the Euler angles θ, φ, and ψ, which determine the orientation of the trihedral $\varepsilon_1, \varepsilon_2, \varepsilon_3$ relative to the trihedral X, Y, Z.

Hence, (III.1) can also be written as

$$dA = \mathfrak{F}(\varepsilon_1, \varepsilon_2, \varepsilon_3, \theta, \varphi, \psi)\, dxdydz. \tag{III.2}$$

This form of (III.1) emphasizes the dependence of the work expended on the deformation of an elementary parallelopiped both on the magnitudes of the principal strain components and on the direction of the fibers of the body which are subjected to these strains.

Hence, (III.2), as well as the equivalent formula (III.1), assumes that the body reacts to deformations differently in different directions, i. e., it assumes that the material of the body is anisotropic. If the physical properties of the body were the same in all directions, the work expended in deforming a volume element would not depend on quantities which vary with a rotation of the coordinate axes, but would be a function only of the invariant quantities.

It follows that for an isotropic body

$$dA = \mathfrak{F}(\varepsilon_1, \varepsilon_2, \varepsilon_3)\, dxdydz. \tag{III.3}$$

The three independent invariants $\varepsilon_1, \varepsilon_2, \varepsilon_3$ are of value because they have a simple physical meaning, especially for small deformations. Mathematically, however, they are inconvenient because, in order to express them in terms of the strain components, the cubic equation (I.86) would have to be solved.

In view of this, it is more expedient to express the work of deformation on an element of an isotropic body as a function of the three coefficients of equation (I.86)

$$a_2 = \varepsilon_{xx} + \varepsilon_{yy} + \varepsilon_{zz} = \varepsilon_1 + \varepsilon_2 + \varepsilon_3,$$

$$a_1 = \varepsilon_{xx}\varepsilon_{yy} + \varepsilon_{xx}\varepsilon_{zz} + \varepsilon_{yy}\varepsilon_{zz} - \frac{1}{4}(\varepsilon_{xy}^2 + \varepsilon_{xz}^2 + \varepsilon_{yz}^2) = \varepsilon_1\varepsilon_2 + \varepsilon_1\varepsilon_3 + \varepsilon_2\varepsilon_3, \qquad \text{(III.4)}$$

$$a_0 = \varepsilon_{xx}\varepsilon_{yy}\varepsilon_{zz} - \frac{1}{4}(\varepsilon_{xx}\varepsilon_{yz}^2 + \varepsilon_{yy}\varepsilon_{xz}^2 + \varepsilon_{zz}\varepsilon_{xy}^2 - \varepsilon_{xy}\varepsilon_{xz}\varepsilon_{yz}),$$

rather than in terms of the roots of this equation. These coefficients have the advantage over $\varepsilon_1, \varepsilon_2, \varepsilon_3$ of being rational functions of the strain components.

Then the work done in deforming an elementary parallelopiped of an isotropic body is most conveniently written in the form

$$dA = \Phi(a_2, a_1, a_0)\, dxdydz. \qquad \text{(III.5)}$$

It follows that the work done in deforming the whole body is

$$A = \iiint \Phi(a_2, a_1, a_0)\, dxdydz, \qquad \text{(III.6)}$$

where the integration must be extended over the whole volume of the body in its unstrained state, since $dxdydz$ is the volume of an elementary parallelopiped extracted from the body in its initial state.

The function $\Phi(a_2, a_1, a_0)$ can be called the work of deformation or the strain energy referred to a unit volume of the body in its unstrained state or, more briefly, the specific strain energy.

§ 26. The Principle of Virtual Displacements

Let us apply the principle of virtual displacements to a deformed body in a state of equilibrium. To this end, assign to the displacements $u(x, y, z)$, $v(x, y, z)$, $w(x, y, z)$ virtual increments δu, δv, δw, respectively, which will be regarded as arbitrary continuous functions of x, y, z, equal to zero at those points where the values of the displacements are given.

Then the strain energy changes by the amount δA, and this must be equal to the work done by all the exterior forces applied to the body in effecting the above virtual displacements. Hence we obtain the equation

$$\delta A = \delta R_1 + \delta R_2, \tag{III.7}$$

where δR_1 is the virtual work due to the body forces and δR_2 is the virtual work of the surface forces.

The virtual work of the body forces is equal to

$$\delta R_1 = \iiint [F_\xi^* \delta u + F_\eta^* \delta v + F_\zeta^* \delta w]\, D dx dy dz, \tag{III.8}$$

where, as before, F_ξ^*, F_η^*, F_ζ^* are the components along the X-, Y-, Z-axes of the force referred to a unit volume of the deformed body, $dxdydz$ is the volume element of the unstrained body, and $Ddxdydz = \frac{V^*}{V} dxdydz$ is the corresponding volume element of the strained body. Here again, regarding x, y, z as the independent variables, the integration in (III.8) must be extended over the body in its initial state.

The virtual work of the surface forces is equal to

$$\delta R_2 = \iint [f_\xi^* \delta u + f_\eta^* \delta v + f_\zeta^* \delta w]\, d\Omega', \tag{III.9}$$

where f_ξ^*, f_η^*, f_ζ^* are the components along the X-, Y-, Z-axes of the force acting on a unit area of the surface of the deformed body, while $d\Omega'$ is the differential of this area.

In order to carry out the integration in (III.9) over the surface of the body in its initial state rather than in its strained state, let us express $d\Omega'$ in terms of $d\Omega$, the differential of surface area in the unstrained state.

To this end, let us consider an elementary rectangular area with sides $d\mathbf{a}$, $d\mathbf{b}$, on the bounding surface of the body before deformation. The elementary rectangle has an area equal to

$$d\Omega = da\,db,$$

and its orientation is determined by the direction of the unit normal vector

$$\mathbf{i}_n = \frac{d\mathbf{a} \times d\mathbf{b}}{d\Omega}. \qquad \text{(III.10)}$$

As a result of the deformation, this rectangular area is transformed into a surface element of the deformed body having the form of a parallelogram with sides $(1+E_a)da$, $(1+E_b)db$. The cosine of the angle formed by these sides is

$$\cos(d\mathbf{a}',\ d\mathbf{b}') = \frac{\varepsilon_{ab}}{(1+E_a)(1+E_b)} = \frac{\varepsilon_{ab}}{\sqrt{(1+2\varepsilon_{aa})(1+2\varepsilon_{bb})}}, \qquad \text{(III.11)}$$

where E_a, E_b are the relative elongations in the directions da, db and ε_{aa}, ε_{bb}, ε_{ab} are the corresponding strain components.

Hence an element on the surface of the strained body has an area equal to

$$d\Omega' = \sqrt{(1+2\varepsilon_{aa})(1+2\varepsilon_{bb}) - \varepsilon_{ab}^2} \cdot d\Omega = \frac{S_n^*}{S_n}\,d\Omega. \qquad \text{(III.12)}$$

The coefficient $\frac{S_n^*}{S_n}$, which is equal to the ratio of the elements of area in the terminal and initial states, can be expressed in terms of the strain components relative to the X-, Y-, Z-axes and the three direction cosines of the normal to the surface in the unstrained state. To do this, it is necessary to replace the strain components

$\varepsilon_{aa}, \varepsilon_{bb}, \varepsilon_{ab}$ in the formula

$$\frac{S_n^*}{S_n} = \sqrt{(1+2\varepsilon_{aa})(1+2\varepsilon_{bb}) - \varepsilon_{ab}^2} \qquad \text{(III.13)}$$

by their expressions in terms of $\varepsilon_{xx}, \ldots, \varepsilon_{yz}$ and the direction cosines of $d\mathbf{a}$, $d\mathbf{b}$, and $\mathbf{n}$.

Then, after some elementary, but involved, transformations (using (I.36), (I.33), and (I.59)), we have

$$\frac{S_n^*}{S_n} = \sqrt{\gamma_{xx}\cos^2(n, x) + \gamma_{yy}\cos^2(n, y) + \gamma_{zz}\cos^2(n, z) +}$$
$$\longrightarrow \overline{+\gamma_{xy}\cos(n, x)\cos(n, y) +}$$
$$\rightarrow \overline{+\gamma_{xz}\cos(n, x)\cos(n, z) + \gamma_{yz}\cos(n, y)\cos(n, z)}, \qquad \text{(III.14)}$$

where

$$\begin{aligned} &\gamma_{xx} = \left(\frac{S_x^*}{S_x}\right)^2, \quad \gamma_{yy} = \left(\frac{S_y^*}{S_y}\right)^2, \quad \gamma_{zz} = \left(\frac{S_z^*}{S_z}\right)^2, \\ &\frac{1}{2}\gamma_{xy} = -\varepsilon_{xy} + \varepsilon_{xz}\varepsilon_{yz} - 2\varepsilon_{xy}\varepsilon_{zz}, \\ &\frac{1}{2}\gamma_{xz} = -\varepsilon_{xz} + \varepsilon_{xy}\varepsilon_{yz} - 2\varepsilon_{xz}\varepsilon_{yy}, \\ &\frac{1}{2}\gamma_{yz} = -\varepsilon_{yz} + \varepsilon_{xz}\varepsilon_{xy} - 2\varepsilon_{yz}\varepsilon_{xx}, \end{aligned} \qquad \text{(III.15)}$$

and $\widehat{nx}$, $\widehat{ny}$, $\widehat{nz}$ are the direction angles of the normal to the bounding surface in the unstrained state relative to the X-, Y-, Z-axes.

Hence, the virtual work done by the external surface forces through the virtual displacements $\delta u, \delta v, \delta w$ is of the form

$$\delta R_2 = \iint [f_\xi^* \delta u + f_\eta^* \delta v + f_\zeta^* \delta w] \frac{S_n^*}{S_n}\, d\Omega, \qquad \text{(III.16)}$$

where $d\Omega$ is the area of a surface element in the initial state. Hence the integration in (III.16) must be extended over the surface in the unstrained state, rather than the surface in the terminal state as was the case in (III.9).

The fact that all the volume and surface integrals appearing in (III.7) are now to be extended over the limits of the body in the unstrained state (and not in the strained state) is a great convenience, since the limits of integration are now independent of any unknown quantities (translations).

§ 27. Derivation of the Differential Equations of Equilibrium of a Deformed Isotropic Body from the Principle of Virtual Displacements

The principle formulated in the preceding section enables us to obtain the differential equations governing the equilibrium of a body in a state of strain. Let us derive these equations on the assumption that the strain energy is given by (III.6), i. e., on the assumption that the body is homogeneous and isotropic and that the dissipative forces play a negligible role in the deformation.

Then

$$\delta A = \delta \int\int\int \Phi(a_2, a_1, a_0)\, dxdydz$$

$$= \int\int\int \delta\,[\Phi(a_2, a_1, a_0)]\, dxdydz. \qquad \text{(III.17)}$$

On the other hand,

$$\delta\,[\Phi(a_2, a_1, a_0)] = \frac{\partial\Phi}{\partial\varepsilon_{xx}}\,\delta\varepsilon_{xx} + \frac{\partial\Phi}{\partial\varepsilon_{yy}}\,\delta\varepsilon_{yy} + \frac{\partial\Phi}{\partial\varepsilon_{zz}}\,\delta\varepsilon_{zz}$$

$$+ \frac{\partial\Phi}{\partial\varepsilon_{xy}}\,\delta\varepsilon_{xy} + \frac{\partial\Phi}{\partial\varepsilon_{xz}}\,\delta\varepsilon_{xz} + \frac{\partial\Phi}{\partial\varepsilon_{yz}}\,\delta\varepsilon_{yz}, \qquad \text{(III.18)}$$

where, in accordance with equations (III.4),

$$\frac{\partial\Phi}{\partial\varepsilon_{xx}} = \frac{\partial\Phi}{\partial a_2}\frac{\partial a_2}{\partial\varepsilon_{xx}} + \frac{\partial\Phi}{\partial a_1}\frac{\partial a_1}{\partial\varepsilon_{xx}} + \frac{\partial\Phi}{\partial a_0}\frac{\partial a_0}{\partial\varepsilon_{xx}}$$

$$= \frac{\partial\Phi}{\partial a_2} + \frac{\partial\Phi}{\partial a_1}(\varepsilon_{yy} + \varepsilon_{zz}) + \frac{\partial\Phi}{\partial a_0}\left(\varepsilon_{yy}\varepsilon_{zz} - \frac{1}{4}\varepsilon_{yz}^2\right),$$

$$\frac{\partial\Phi}{\partial\varepsilon_{yy}}=\frac{\partial\Phi}{\partial a_2}+\frac{\partial\Phi}{\partial a_1}(\varepsilon_{xx}+\varepsilon_{zz})+\frac{\partial\Phi}{\partial a_0}\left(\varepsilon_{xx}\varepsilon_{zz}-\frac{1}{4}\varepsilon_{xz}^2\right),$$

$$\frac{\partial\Phi}{\partial\varepsilon_{zz}}=\frac{\partial\Phi}{\partial a_2}+\frac{\partial\Phi}{\partial a_1}(\varepsilon_{xx}+\varepsilon_{yy})+\frac{\partial\Phi}{\partial a_0}\left(\varepsilon_{xx}\varepsilon_{yy}-\frac{1}{4}\varepsilon_{yx}^2\right), \quad \text{(III.19)}$$

$$\frac{\partial\Phi}{\partial\varepsilon_{xy}}=\frac{\partial\Phi}{\partial a_2}\frac{\partial a_2}{\partial\varepsilon_{xy}}+\frac{\partial\Phi}{\partial a_1}\frac{\partial a_1}{\partial\varepsilon_{xy}}+\frac{\partial\Phi}{\partial a_0}\frac{\partial a_0}{\partial\varepsilon_{xy}}$$

$$=-\frac{1}{2}\frac{\partial\Phi}{\partial a_1}\varepsilon_{xy}+\frac{1}{2}\frac{\partial\Phi}{\partial a_0}\left(\frac{1}{2}\varepsilon_{xz}\varepsilon_{yz}-\varepsilon_{xy}\varepsilon_{zz}\right),$$

$$\frac{\partial\Phi}{\partial\varepsilon_{xz}}=-\frac{1}{2}\frac{\partial\Phi}{\partial a_1}\varepsilon_{xz}+\frac{1}{2}\frac{\partial\Phi}{\partial a_0}\left(\frac{1}{2}\varepsilon_{xy}\varepsilon_{yz}-\varepsilon_{xz}\varepsilon_{yy}\right),$$

$$\frac{\partial\Phi}{\partial\varepsilon_{yz}}=-\frac{1}{2}\frac{\partial\Phi}{\partial a_1}\varepsilon_{yz}+\frac{1}{2}\frac{\partial\Phi}{\partial a_0}\left(\frac{1}{2}\varepsilon_{xy}\varepsilon_{xz}-\varepsilon_{yz}\varepsilon_{xx}\right).$$

Moreover, taking into account that, by (I.22),

$$\delta(\varepsilon_{xx})=\left(1+\frac{\partial u}{\partial x}\right)\delta\left(\frac{\partial u}{\partial x}\right)+\frac{\partial v}{\partial x}\delta\left(\frac{\partial v}{\partial x}\right)+\frac{\partial w}{\partial x}\delta\left(\frac{\partial w}{\partial x}\right)$$

$$=(1+e_{xx})\frac{\partial(\delta u)}{\partial x}+\left(\frac{1}{2}e_{xy}+\omega_z\right)\frac{\partial(\delta v)}{\partial x}$$

$$+\left(\frac{1}{2}e_{xz}-\omega_y\right)\frac{\partial(\delta w)}{\partial x},$$

$$\delta(\varepsilon_{yy})=\left(\frac{1}{2}e_{xy}-\omega_z\right)\frac{\partial(\delta u)}{\partial y}+(1+e_{yy})\frac{\partial(\delta v)}{\partial y}$$

$$+\left(\frac{1}{2}e_{yz}+\omega_x\right)\frac{\partial(\delta w)}{\partial y},$$

$$\delta(\varepsilon_{zz})=\left(\frac{1}{2}e_{xz}+\omega_y\right)\frac{\partial(\delta u)}{\partial z}+\left(\frac{1}{2}e_{yz}-\omega_x\right)\frac{\partial(\delta v)}{\partial z}$$

$$+(1+e_{zz})\frac{\partial(\delta w)}{\partial z}, \quad \text{(III.20)}$$

$$\delta(\varepsilon_{xy})=\frac{\partial u}{\partial y}\delta\left(\frac{\partial u}{\partial x}\right)+\left(1+\frac{\partial v}{\partial y}\right)\delta\left(\frac{\partial v}{\partial x}\right)$$

$$+\frac{\partial w}{\partial y}\delta\left(\frac{\partial w}{\partial x}\right)+\left(1+\frac{\partial u}{\partial x}\right)\delta\left(\frac{\partial u}{\partial y}\right)+\frac{\partial v}{\partial x}\delta\left(\frac{\partial v}{\partial y}\right)$$

$$+\frac{\partial w}{\partial x}\delta\left(\frac{\partial w}{\partial y}\right)=\left(\frac{1}{2}e_{xy}-\omega_z\right)\frac{\partial(\delta u)}{\partial x}$$

$$+(1+e_{yy})\frac{\partial(\delta v)}{\partial x}+\left(\frac{1}{2}e_{yz}+\omega_x\right)\frac{\partial(\delta w)}{\partial x}$$

$$+(1+e_{xx})\frac{\partial(\delta u)}{\partial y}+\left(\frac{1}{2}e_{xy}+\omega_z\right)\frac{\partial(\delta v)}{\partial y}$$

$$+\left(\frac{1}{2}e_{xz}-\omega_y\right)\frac{\partial(\delta w)}{\partial y},$$

$$\delta(\varepsilon_{xz})=\left(\frac{1}{2}e_{xz}+\omega_y\right)\frac{\partial(\delta u)}{\partial x}+\left(\frac{1}{2}e_{yz}-\omega_x\right)\frac{\partial(\delta v)}{\partial x}$$

$$+(1+e_{zz})\frac{\partial(\delta w)}{\partial x}+(1+e_{xx})\frac{\partial(\delta u)}{\partial z}$$

$$+\left(\frac{1}{2}e_{xy}+\omega_z\right)\frac{\partial(\delta v)}{\partial z}+\left(\frac{1}{2}e_{xz}-\omega_y\right)\frac{\partial(\delta w)}{\partial z},$$

$$\delta(\varepsilon_{yz})=\left(\frac{1}{2}e_{xz}+\omega_y\right)\frac{\partial(\delta u)}{\partial y}+\left(\frac{1}{2}e_{yz}-\omega_x\right)\frac{\partial(\delta v)}{\partial y}$$

$$+(1+e_{zz})\frac{\partial(\delta w)}{\partial y}+\left(\frac{1}{2}e_{xy}-\omega_z\right)\frac{\partial(\delta u)}{\partial z}$$

$$+(1+e_{yy})\frac{\partial(\delta v)}{\partial z}+\left(\frac{1}{2}e_{yz}+\omega_x\right)\frac{\partial(\delta w)}{\partial z},$$

and substituting these values of $\delta(\varepsilon_{xx}), \ldots, \delta(\varepsilon_{yz})$ into (III.18), we obtain

$$\delta[\Phi(a_2, a_1, a_0)]=\frac{\partial\Phi}{\partial u_x}\frac{\partial(\delta u)}{\partial x}+\frac{\partial\Phi}{\partial u_y}\frac{\partial(\delta u)}{\partial y}+\frac{\partial\Phi}{\partial u_z}\frac{\partial(\delta u)}{\partial z}$$

$$+\frac{\partial\Phi}{\partial v_x}\frac{\partial(\delta v)}{\partial x}+\frac{\partial\Phi}{\partial v_y}\frac{\partial(\delta v)}{\partial y}+\frac{\partial\Phi}{\partial v_z}\frac{\partial(\delta v)}{\partial z}$$

$$+\frac{\partial\Phi}{\partial w_x}\frac{\partial(\delta w)}{\partial x}+\frac{\partial\Phi}{\partial w_y}\frac{\partial(\delta w)}{\partial y}+\frac{\partial\Phi}{\partial w_z}\frac{\partial(\delta w)}{\partial z}, \qquad \text{(III.21)}$$

where

$$\frac{\partial\Phi}{\partial u_x}=\frac{\partial\Phi}{\partial\left(\frac{\partial u}{\partial x}\right)}=(1+e_{xx})\frac{\partial\Phi}{\partial\varepsilon_{xx}}+\left(\frac{1}{2}e_{xy}-\omega_z\right)\frac{\partial\Phi}{\partial\varepsilon_{xy}}$$

$$+\left(\frac{1}{2}e_{xz}+\omega_y\right)\frac{\partial\Phi}{\partial\varepsilon_{xz}};$$

$$\frac{\partial\Phi}{\partial u_y}=\frac{\partial\Phi}{\partial\left(\frac{\partial u}{\partial y}\right)}=(1+e_{xx})\frac{\partial\Phi}{\partial\varepsilon_{xy}}+\left(\frac{1}{2}e_{xy}-\omega_z\right)\frac{\partial\Phi}{\partial\varepsilon_{yy}}$$

$$+\left(\frac{1}{2}e_{xz}+\omega_y\right)\frac{\partial\Phi}{\partial\varepsilon_{yz}},$$

$$\frac{\partial\Phi}{\partial u_z}=\frac{\partial\Phi}{\partial\left(\frac{\partial u}{\partial z}\right)}=(1+e_{xx})\frac{\partial\Phi}{\partial\varepsilon_{xz}}+\left(\frac{1}{2}e_{xy}-\omega_z\right)\frac{\partial\Phi}{\partial\varepsilon_{yz}}$$

$$+\left(\frac{1}{2}e_{xz}+\omega_y\right)\frac{\partial\Phi}{\partial\varepsilon_{zz}},$$

$$\frac{\partial\Phi}{\partial v_x}=\frac{\partial\Phi}{\partial\left(\frac{\partial v}{\partial x}\right)}=\left(\frac{1}{2}e_{xy}+\omega_z\right)\frac{\partial\Phi}{\partial\varepsilon_{xx}}+(1+e_{yy})\frac{\partial\Phi}{\partial\varepsilon_{xy}}$$

$$+\left(\frac{1}{2}e_{yz}-\omega_x\right)\frac{\partial\Phi}{\partial\varepsilon_{xz}}; \quad \text{(III.22)}$$

$$\frac{\partial\Phi}{\partial v_y}=\frac{\partial\Phi}{\partial\left(\frac{\partial v}{\partial y}\right)}=\left(\frac{1}{2}e_{xy}+\omega_z\right)\frac{\partial\Phi}{\partial\varepsilon_{xy}}+(1+e_{yy})\frac{\partial\Phi}{\partial\varepsilon_{yy}}$$

$$+\left(\frac{1}{2}e_{yz}-\omega_x\right)\frac{\partial\Phi}{\partial\varepsilon_{yz}},$$

$$\frac{\partial\Phi}{\partial v_z}=\frac{\partial\Phi}{\partial\left(\frac{\partial v}{\partial z}\right)}=\left(\frac{1}{2}e_{xy}+\omega_z\right)\frac{\partial\Phi}{\partial\varepsilon_{xz}}+(1+e_{yy})\frac{\partial\Phi}{\partial\varepsilon_{yz}}$$

$$+\left(\frac{1}{2}e_{yz}-\omega_x\right)\frac{\partial\Phi}{\partial\varepsilon_{zz}},$$

$$\frac{\partial\Phi}{\partial w_x}=\frac{\partial\Phi}{\partial\left(\frac{\partial w}{\partial x}\right)}=\left(\frac{1}{2}e_{xz}-\omega_y\right)\frac{\partial\Phi}{\partial\varepsilon_{xx}}+\left(\frac{1}{2}e_{yz}+\omega_x\right)\frac{\partial\Phi}{\partial\varepsilon_{xy}}$$

$$+(1+e_{zz})\frac{\partial\Phi}{\partial\varepsilon_{xz}},$$

$$\frac{\partial\Phi}{\partial w_y}=\frac{\partial\Phi}{\partial\left(\frac{\partial w}{\partial y}\right)}=\left(\frac{1}{2}e_{xz}-\omega_y\right)\frac{\partial\Phi}{\partial\varepsilon_{xy}}+\left(\frac{1}{2}e_{yz}+\omega_x\right)\frac{\partial\Phi}{\partial\varepsilon_{yy}}$$

$$+(1+e_{zz})\frac{\partial\Phi}{\partial\varepsilon_{yz}},$$

$$\frac{\partial\Phi}{\partial w_z}=\frac{\partial\Phi}{\partial\left(\frac{\partial w}{\partial z}\right)}=\left(\frac{1}{2}e_{xz}-\omega_y\right)\frac{\partial\Phi}{\partial\varepsilon_{xz}}+\left(\frac{1}{2}e_{yz}+\omega_x\right)\frac{\partial\Phi}{\partial\varepsilon_{yz}}$$

$$+(1+e_{zz})\frac{\partial\Phi}{\partial\varepsilon_{zz}}.$$

Here $\frac{\partial\Phi}{\partial\varepsilon_{xx}}, \ldots, \frac{\partial\Phi}{\partial\varepsilon_{zz}}$ are determined from (III.19).

Denoting now by (III.23)

U the vector with the projections $\frac{\partial\Phi}{\partial u_x}, \frac{\partial\Phi}{\partial u_y}, \frac{\partial\Phi}{\partial u_z}$ on the X-, Y-, Z-axes,

V the vector with the projections $\frac{\partial\Phi}{\partial v_x}, \frac{\partial\Phi}{\partial v_y}, \frac{\partial\Phi}{\partial v_z}$ on the X-, Y-, Z-axes,

W the vector with the projections $\frac{\partial\Phi}{\partial w_x}, \frac{\partial\Phi}{\partial w_y}, \frac{\partial\Phi}{\partial w_z}$ on the X-, Y-, Z-axes,

we can write (III.21) more briefly as

$$\delta[\Phi(a_2, a_1, a_0)] = \mathbf{U}\cdot\operatorname{grad}(\delta u) + \mathbf{V}\cdot\operatorname{grad}(\delta v) + \mathbf{W}\cdot\operatorname{grad}(\delta w), \tag{III.24}$$

where the symbol "·" denotes the operation of scalar multiplication of two vectors and the symbol "grad" denotes the gradient.

Substituting this result in (III.17) we obtain

$$\delta A = \iiint [\mathbf{U}\cdot\operatorname{grad}(\delta u) + \mathbf{V}\cdot\operatorname{grad}(\delta v) + \mathbf{W}\cdot\operatorname{grad}(\delta w)]\, dxdydz. \tag{III.25}$$

We observe, moreover, that the relation

$$\operatorname{div}(\varphi\boldsymbol{\psi}) = \frac{\partial(\varphi\psi_x)}{\partial x} + \frac{\partial(\varphi\psi_y)}{\partial y} + \frac{\partial(\varphi\psi_z)}{\partial z} = \left(\frac{\partial\psi_x}{\partial x} + \frac{\partial\psi_y}{\partial y} + \frac{\partial\psi_z}{\partial z}\right)\varphi + \psi_x\frac{\partial\varphi}{\partial x} + \psi_y\frac{\partial\varphi}{\partial y} + \psi_z\frac{\partial\varphi}{\partial z} = \varphi\operatorname{div}\boldsymbol{\psi} + \boldsymbol{\psi}\cdot\operatorname{grad}\varphi \tag{III.26}$$

holds for any scalar function $\varphi(x,y,z)$ and any vector function $\boldsymbol{\psi}(x,y,z)$.

In view of this identity we can write

$$\begin{aligned} \mathbf{U}\cdot\operatorname{grad}(\delta u) &= \operatorname{div}(\mathbf{U}\delta u) - \delta u \operatorname{div}\mathbf{U}, \\ \mathbf{V}\cdot\operatorname{grad}(\delta v) &= \operatorname{div}(\mathbf{V}\delta v) - \delta v \operatorname{div}\mathbf{V}, \\ \mathbf{W}\cdot\operatorname{grad}(\delta w) &= \operatorname{div}(\mathbf{W}\delta w) - \delta w \operatorname{div}\mathbf{W}. \end{aligned} \tag{III.27}$$

Accordingly, (III.25) may be rewritten in the form

$$\delta A = \iiint [\operatorname{div}(\mathbf{U}\delta u + \mathbf{V}\delta v + \mathbf{W}\delta w)]\, dxdydz - \iiint [\delta u \operatorname{div}\mathbf{U} + \delta v \operatorname{div}\mathbf{V} + \delta w \operatorname{div}\mathbf{W}]\, dxdydz. \qquad \text{(III.28)}$$

Now, the volume integral of the divergence of a vector can be reduced by Gauss' theorem to a surface integral, namely

$$\iiint [\operatorname{div}\mathbf{Q}]\, dxdydz = \iint Q_n d\Omega, \qquad \text{(III.29)}$$

where Q_n is the projection of the vector Q on the exterior normal to the bounding surface (see, e. g., *Vector Analysis* by N. E. Kochin, p. 144; or *A Course of Theoretical Mechanics* by Y. I. Frenkel, p. 85. An alternative reference in English is A. P. Wills, *Vector Analysis with an Introduction to Tensor Analysis*, New York,1931,pp.101-102. (Trans.)).

In view of this, the formula for the virtual work given in (III.25) may be expressed in the following final form:

$$\delta A = -\iiint [\delta u \operatorname{div}\mathbf{U} + \delta v \operatorname{div}\mathbf{V} + \delta w \operatorname{div}\mathbf{W}]\, dxdydz + \iint [U_n\delta u + V_n\delta v + W_n\delta w]\, d\Omega\,. \qquad \text{(III.30)}$$

Here the first integral is taken over the whole body in its initial state and the second integral is extended over the entire bounding surface in the unstrained state, while U_n, V_n, W_n are the projections of the vectors U, V, W on the exterior normal to the surface.

Substituting (III.30), (III.8), and (III.16) into (III.7), and transposing all the resulting terms to the right-hand side, we obtain

$$\iiint\left\{\left(\operatorname{div}\mathbf{U}+DF_{\xi}^{*}\right)\delta u+\left(\operatorname{div}\mathbf{V}+DF_{\eta}^{*}\right)\delta v\right.$$
$$\left.+\left(\operatorname{div}\mathbf{W}+DF_{\zeta}^{*}\right)\delta w\right\}dx\,dy\,dz$$
$$+\iint\left\{\left(\frac{S_n^*}{S_n}f_{\xi}^{*}-U_n\right)\delta u+\left(\frac{S_n^*}{S_n}f_{\eta}^{*}-V_n\right)\delta v\right. \tag{III.31}$$
$$\left.+\left(\frac{S_n^*}{S_n}f_{\zeta}^{*}-W_n\right)\delta w\right\}d\Omega=0.$$

Since, in accordance with the principle of virtual displacements, this equation must be satisfied for arbitrary values of δu, δv, δw, we obtain the following three equations which must hold at all interior points of the body:

$$\begin{aligned}&\operatorname{div}\mathbf{U}+F_{\xi}^{*}D=0,\\&\operatorname{div}\mathbf{V}+F_{\eta}^{*}D=0,\\&\operatorname{div}\mathbf{W}+F_{\zeta}^{*}D=0,\end{aligned} \tag{III.32}$$

together with the three equations

$$U_n=\frac{S_n^*}{S_n}f_{\xi}^{*},\quad V_n=\frac{S_n^*}{S_n}f_{\eta}^{*},\quad W_n=\frac{S_n^*}{S_n}f_{\zeta}^{*}, \tag{III.33}$$

which must be satisfied at all those points of the bounding surface at which not the displacements, but the forces, are prescribed (at those points where the displacements are prescribed the virtual displacements δu, δv, δw, as noted above, are assumed to be zero, and therefore on such regions the terms in the second integral of (III.31) will vanish). Taking into account that the projections of the vectors **U**, **V**, **W** are given in (III.23), we can rewrite (III.32) in the following form:

$$\frac{\partial}{\partial x}\left(\frac{\partial\Phi}{\partial u_x}\right)+\frac{\partial}{\partial y}\left(\frac{\partial\Phi}{\partial u_y}\right)+\frac{\partial}{\partial z}\left(\frac{\partial\Phi}{\partial u_z}\right)+DF_{\xi}^{*}=0,$$

$$\frac{\partial}{\partial x}\left(\frac{\partial\Phi}{\partial v_x}\right)+\frac{\partial}{\partial y}\left(\frac{\partial\Phi}{\partial v_y}\right)+\frac{\partial}{\partial z}\left(\frac{\partial\Phi}{\partial v_z}\right)+DF^*_\eta=0, \quad \text{(III.34)}$$

$$\frac{\partial}{\partial x}\left(\frac{\partial\Phi}{\partial w_x}\right)+\frac{\partial}{\partial y}\left(\frac{\partial\Phi}{\partial w_y}\right)+\frac{\partial}{\partial z}\left(\frac{\partial\Phi}{\partial w_z}\right)+DF^*_\zeta=0.$$

Or, if $\frac{\partial\Phi}{\partial u_x}, \ldots, \frac{\partial\Phi}{\partial w_z}$ are replaced by their expressions in (III.22), we obtain

$$\frac{\partial}{\partial x}\left[(1+e_{xx})\frac{\partial\Phi}{\partial\varepsilon_{xx}}+\left(\frac{1}{2}e_{xy}-\omega_z\right)\frac{\partial\Phi}{\partial\varepsilon_{xy}}+\left(\frac{1}{2}e_{xz}+\omega_y\right)\frac{\partial\Phi}{\partial\varepsilon_{xz}}\right]$$
$$+\frac{\partial}{\partial y}\left[(1+e_{xx})\frac{\partial\Phi}{\partial\varepsilon_{xy}}+\left(\frac{1}{2}e_{xy}-\omega_z\right)\frac{\partial\Phi}{\partial\varepsilon_{yy}}\right.$$
$$\left.+\left(\frac{1}{2}e_{xz}+\omega_y\right)\frac{\partial\Phi}{\partial\varepsilon_{yz}}\right]+\frac{\partial}{\partial z}\left[(1+e_{xx})\frac{\partial\Phi}{\partial\varepsilon_{xz}}\right.$$
$$\left.+\left(\frac{1}{2}e_{xy}-\omega_z\right)\frac{\partial\Phi}{\partial\varepsilon_{yz}}+\left(\frac{1}{2}e_{xz}+\omega_y\right)\frac{\partial\Phi}{\partial\varepsilon_{zz}}\right]+\frac{V^*}{V}F^*_\xi=0,$$

$$\frac{\partial}{\partial x}\left[\left(\frac{1}{2}e_{xy}+\omega_z\right)\frac{\partial\Phi}{\partial\varepsilon_{xx}}+(1+e_{yy})\frac{\partial\Phi}{\partial\varepsilon_{xy}}+\left(\frac{1}{2}e_{yz}-\omega_x\right)\frac{\partial\Phi}{\partial\varepsilon_{xz}}\right]$$
$$+\frac{\partial}{\partial y}\left[\left(\frac{1}{2}e_{xy}+\omega_z\right)\frac{\partial\Phi}{\partial\varepsilon_{xy}}+(1+e_{yy})\frac{\partial\Phi}{\partial\varepsilon_{yy}}\right.$$
$$\left.+\left(\frac{1}{2}e_{yz}-\omega_x\right)\frac{\partial\Phi}{\partial\varepsilon_{yz}}\right]+ \quad \text{(III.35)}$$
$$+\frac{\partial}{\partial z}\left[\left(\frac{1}{2}e_{xy}+\omega_z\right)\frac{\partial\Phi}{\partial\varepsilon_{xz}}+(1+e_{yy})\frac{\partial\Phi}{\partial\varepsilon_{yz}}\right.$$
$$\left.+\left(\frac{1}{2}e_{yz}-\omega_x\right)\frac{\partial\Phi}{\partial\varepsilon_{zz}}\right]+\frac{V^*}{V}F^*_\eta=0,$$

$$\frac{\partial}{\partial x}\left[\left(\frac{1}{2}e_{xz}-\omega_y\right)\frac{\partial\Phi}{\partial\varepsilon_{xx}}+\left(\frac{1}{2}e_{yz}+\omega_x\right)\frac{\partial\Phi}{\partial\varepsilon_{xy}}+(1+e_{zz})\frac{\partial\Phi}{\partial\varepsilon_{xz}}\right]$$
$$+\frac{\partial}{\partial y}\left[\left(\frac{1}{2}e_{xz}-\omega_y\right)\frac{\partial\Phi}{\partial\varepsilon_{xy}}+\left(\frac{1}{2}e_{yz}+\omega_x\right)\frac{\partial\Phi}{\partial\varepsilon_{yy}}\right.$$
$$\left.+(1+e_{zz})\frac{\partial\Phi}{\partial\varepsilon_{yz}}\right]+\frac{\partial}{\partial z}\left[\left(\frac{1}{2}e_{xz}-\omega_y\right)\frac{\partial\Phi}{\partial\varepsilon_{xz}}+\left(\frac{1}{2}e_{yz}+\omega_x\right)\frac{\partial\Phi}{\partial\varepsilon_{yz}}\right.$$
$$\left.+(1+e_{zz})\frac{\partial\Phi}{\partial\varepsilon_{zz}}\right]+\frac{V^*}{V}F^*_\zeta=0.$$

As soon as the function $\Phi(a_2, a_1, a_0)$ is known, equations (III.35) form a system of three equations in three unknowns, the displacements u, v, w of the points of the body. This reduces the study of the equilibrium of a deformed body to the solution of this system of equations.

§ 28. The Relation between Stress and Strain Components

Let us compare equations (III.35) with equations (III.43), both of which express the conditions of equilibrium of a volume element of the deformed body which initially was a rectangular parallelopiped with edges dx, dy, dz parallel to the X-, Y-, Z-axes. It is seen that one of these systems is transformed into the other by setting

$$\begin{aligned} \sigma^*_{xx} &= \frac{\partial\Phi}{\partial\varepsilon_{xx}}, & \sigma^*_{xy} &= \frac{\partial\Phi}{\partial\varepsilon_{xy}}, & \sigma^*_{xz} &= \frac{\partial\Phi}{\partial\varepsilon_{xz}}, \\ \sigma^*_{yx} &= \frac{\partial\Phi}{\partial\varepsilon_{xy}}, & \sigma^*_{yy} &= \frac{\partial\Phi}{\partial\varepsilon_{yy}}, & \sigma^*_{yz} &= \frac{\partial\Phi}{\partial\varepsilon_{yz}}, \\ \sigma^*_{zx} &= \frac{\partial\Phi}{\partial\varepsilon_{xz}}, & \sigma^*_{zy} &= \frac{\partial\Phi}{\partial\varepsilon_{yz}}, & \sigma^*_{zz} &= \frac{\partial\Phi}{\partial\varepsilon_{zz}}. \end{aligned} \qquad \text{(III.36)}$$

Thus, the differential equations governing the equilibrium of a body in a state of strain, derived in this chapter from the principle of virtual displacements, will be brought into agreement with the equations of the preceding chapter, which expressed the equilibrium conditions for an infinitesimal element of the body. This will be done by assuming relations of the form (III.36) between the derivatives of the function Φ (the specific strain energy referred to a unit volume of the unstrained body) with respect to the strain components and the quantities σ^*_{ij}.

It follows immediately from (III.36) that

$$\sigma^*_{xy} = \sigma^*_{yx},$$
$$\sigma^*_{xz} = \sigma^*_{zx}, \qquad \text{(III.37)}$$
$$\sigma^*_{yz} = \sigma^*_{zy}.$$

We have already referred to the existence of these relations (Ch. II, § 20), and they are now confirmed. Substituting in equations (III.36) the expressions for σ^*_{ij} in terms of the stress components in the directions i_1, i_2, i_3 given by (II.44), and the expressions for the derivatives of the function $\Phi(a_2, a_1, a_0)$ with respect to the strain components given in (III.19), we obtain

$$\sigma^*_{xx} = \frac{S^*_x}{S_x}\frac{\sigma_{xx}}{1+E_x} = \frac{\partial\Phi}{\partial a_2} + \frac{\partial\Phi}{\partial a_1}(\varepsilon_{yy} + \varepsilon_{zz}) + \frac{\partial\Phi}{\partial a_0}\left(\varepsilon_{yy}\varepsilon_{zz} - \frac{1}{4}\varepsilon^2_{yz}\right),$$

$$\sigma^*_{yy} = \frac{S^*_y}{S_y}\frac{\sigma_{yy}}{1+E_y} = \frac{\partial\Phi}{\partial a_2} + \frac{\partial\Phi}{\partial a_1}(\varepsilon_{xx} + \varepsilon_{zz}) + \frac{\partial\Phi}{\partial a_0}\left(\varepsilon_{xx}\varepsilon_{zz} - \frac{1}{4}\varepsilon^2_{xz}\right),$$

$$\sigma^*_{zz} = \frac{S^*_z}{S_z}\frac{\sigma_{zz}}{1+E_z} = \frac{\partial\Phi}{\partial a_2} + \frac{\partial\Phi}{\partial a_1}(\varepsilon_{xx} + \varepsilon_{yy}) + \frac{\partial\Phi}{\partial a_0}\left(\varepsilon_{xx}\varepsilon_{yy} - \frac{1}{4}\varepsilon^2_{xy}\right), \qquad \text{(III.38)}$$

$$\sigma^*_{xy} = \sigma^*_{yx} = \frac{S^*_x}{S_x}\frac{\sigma_{xy}}{1+E_y} = \frac{S^*_y}{S_y}\frac{\sigma_{yx}}{1+E_x} = -\frac{1}{2}\frac{\partial\Phi}{\partial a_1}\varepsilon_{xy} + \frac{1}{2}\frac{\partial\Phi}{\partial a_0}\left(\frac{1}{2}\varepsilon_{xz}\varepsilon_{yz} - \varepsilon_{zz}\varepsilon_{xy}\right),$$

$$\sigma^*_{xz} = \sigma^*_{zx} = \frac{S^*_x}{S_x}\frac{\sigma_{xz}}{1+E_z} = \frac{S^*_z}{S_z}\frac{\sigma_{zx}}{1+E_x} = -\frac{1}{2}\frac{\partial\Phi}{\partial a_1}\varepsilon_{xz} + \frac{1}{2}\frac{\partial\Phi}{\partial a_0}\left(\frac{1}{2}\varepsilon_{xy}\varepsilon_{yz} - \varepsilon_{yy}\varepsilon_{xz}\right),$$

$$\sigma^*_{yz} = \sigma^*_{zy} = \frac{S^*_y}{S_y}\frac{\sigma_{yz}}{1+E_z} = \frac{S^*_z}{S_z}\frac{\sigma_{zy}}{1+E_y} = -\frac{1}{2}\frac{\partial\Phi}{\partial a_1}\varepsilon_{yz} + \frac{1}{2}\frac{\partial\Phi}{\partial c_0}\left(\frac{1}{2}\varepsilon_{xy}\varepsilon_{xz} - \varepsilon_{xx}\varepsilon_{yz}\right).$$

Here the ratios $\frac{S^*_x}{S_x}$, $\frac{S^*_y}{S_y}$, $\frac{S^*_z}{S_z}$ are expressed in terms of the strain components in (II.36) and (II.37), and the elongations E_x, E_y, E_z in (I.29).

Equations (III.38) are the general statement of the relation which must subsist between the stress and strain components. In deriving these equations we have used only two assumptions. These assumptions were that the body is isotropic and that the dissipative forces due to the interaction of the particles of the body are small enough to be neglected in comparison with the conservative forces.

The right-hand sides of equations (III.38) are linear functions of the components of three symmetric tensors of the second order

$$\Pi_0 = \begin{pmatrix} 1, & 0, & 0 \\ 0, & 1, & 0 \\ 0, & 0, & 1 \end{pmatrix}, \quad \text{(the unit tensor)}$$

$$\Pi_1 = \begin{pmatrix} \varepsilon_{yy}+\varepsilon_{zz}, & -\frac{1}{2}\varepsilon_{xy}, & -\frac{1}{2}\varepsilon_{xz} \\ -\frac{1}{2}\varepsilon_{xy}, & \varepsilon_{xx}+\varepsilon_{zz}, & -\frac{1}{2}\varepsilon_{yz} \\ -\frac{1}{2}\varepsilon_{xz}, & -\frac{1}{2}\varepsilon_{yz}, & \varepsilon_{xx}+\varepsilon_{yy} \end{pmatrix}, \qquad \text{(III.39)}$$

$$\Pi_2 = \begin{pmatrix} \varepsilon_{yy}\varepsilon_{zz}-\frac{1}{4}\varepsilon^2_{yz}, & \frac{1}{4}\varepsilon_{xz}\varepsilon_{yz}-\frac{1}{2}\varepsilon_{zz}\varepsilon_{xy}, & \frac{1}{4}\varepsilon_{xy}\varepsilon_{yz}-\frac{1}{2}\varepsilon_{yy}\varepsilon_{xz} \\ \frac{1}{4}\varepsilon_{xz}\varepsilon_{yz}-\frac{1}{2}\varepsilon_{zz}\varepsilon_{xy}, & \varepsilon_{xx}\varepsilon_{zz}-\frac{1}{4}\varepsilon^2_{xz}, & \frac{1}{4}\varepsilon_{yx}\varepsilon_{xz}-\frac{1}{2}\varepsilon_{xx}\varepsilon_{yz} \\ \frac{1}{4}\varepsilon_{xy}\varepsilon_{yz}-\frac{1}{2}\varepsilon_{yy}\varepsilon_{xz}, & \frac{1}{4}\varepsilon_{xy}\varepsilon_{xz}-\frac{1}{2}\varepsilon_{xx}\varepsilon_{yz}, & \varepsilon_{xx}\varepsilon_{yy}-\frac{1}{4}\varepsilon^2_{xy} \end{pmatrix},$$

where the coefficients of the components of these

tensors are the three invariants $\frac{\partial\Phi}{\partial a_2}, \frac{\partial\Phi}{\partial a_1}, \frac{\partial\Phi}{\partial a_0}$. Therefore, equations (III.38) can be combined in the matrix relation

$$S = \frac{\partial\Phi}{\partial a_2}\Pi_0 + \frac{\partial\Phi}{\partial a_1}\Pi_1 + \frac{\partial\Phi}{\partial a_0}\Pi_2, \qquad \text{(III.40)}$$

where S denotes the symmetric tensor

$$S = \begin{pmatrix} \overset{*}{\sigma}_{xx}, & \overset{*}{\sigma}_{yx}, & \overset{*}{\sigma}_{zx} \\ \overset{*}{\sigma}_{xy}, & \overset{*}{\sigma}_{yy}, & \overset{*}{\sigma}_{zy} \\ \overset{*}{\sigma}_{xz}, & \overset{*}{\sigma}_{yz}, & \overset{*}{\sigma}_{zz} \end{pmatrix}, \qquad \text{(III.41)}$$

which can be called the generalized stress tensor of an arbitrary deformation. Furthermore, by introducing the notations

$$\begin{aligned} \Psi_2 &= \frac{\partial\Phi}{\partial a_2} + a_2\frac{\partial\Phi}{\partial a_1} + a_1\frac{\partial\Phi}{\partial a_0}, \\ \Psi_1 &= -\frac{\partial\Phi}{\partial a_1} - a_2\frac{\partial\Phi}{\partial a_0}, \\ \Psi_0 &= \frac{\partial\Phi}{\partial a_0}, \end{aligned} \qquad \text{(III.42)}$$

we can reduce equations (III.38) to the form

$$\begin{aligned} \overset{*}{\sigma}_{xx} &= \Psi_2 + \Psi_1\varepsilon_{xx} + \Psi_0\left(\varepsilon_{xx}^2 + \frac{1}{4}\varepsilon_{xy}^2 + \frac{1}{4}\varepsilon_{xz}^2\right), \\ \overset{*}{\sigma}_{yy} &= \Psi_2 + \Psi_1\varepsilon_{yy} + \Psi_0\left(\varepsilon_{yy}^2 + \frac{1}{4}\varepsilon_{xy}^2 + \frac{1}{4}\varepsilon_{yz}^2\right), \\ \overset{*}{\sigma}_{zz} &= \Psi_2 + \Psi_1\varepsilon_{zz} + \Psi_0\left(\varepsilon_{zz}^2 + \frac{1}{4}\varepsilon_{xz}^2 + \frac{1}{4}\varepsilon_{yz}^2\right), \\ \overset{*}{\sigma}_{yx} &= \frac{1}{2}\left\{\Psi_1\varepsilon_{xy} + \Psi_0\left[(\varepsilon_{xx} + \varepsilon_{yy})\varepsilon_{xy} + \frac{1}{2}\varepsilon_{xz}\varepsilon_{yz}\right]\right\}, \\ \overset{*}{\sigma}_{xz} &= \frac{1}{2}\left\{\Psi_1\varepsilon_{xz} + \Psi_0\left[(\varepsilon_{xx} + \varepsilon_{zz})\varepsilon_{xz} + \frac{1}{2}\varepsilon_{xy}\varepsilon_{yz}\right]\right\}, \\ \overset{*}{\sigma}_{yz} &= \frac{1}{2}\left\{\Psi_1\varepsilon_{yz} + \Psi_0\left[(\varepsilon_{yy} + \varepsilon_{zz})\varepsilon_{yz} + \frac{1}{2}\varepsilon_{xy}\varepsilon_{xz}\right]\right\}. \end{aligned} \qquad \text{(III.43)}$$

Hence the above relations between the stresses and strains may be somewhat modified by substituting one set of invariant coefficients for another.

It is essential to observe that in all such transformations these relations always preserve the structure of the equation

$$S = \Psi_2 \Pi_0 + \Psi_1 \Pi_1 + \Psi_0 \Pi_2, \qquad \text{(III.44)}$$

where Ψ_2, Ψ_1, Ψ_0 are functions of the strain invariants Π_0, the unit tensor, Π_1, a tensor whose components are linear combinations of the strain components, and Π_2, a tensor whose components are quadratic combinations of the strain components.

Although the relations (III.38) and (III.43) are somewhat indeterminate since they contain the unknown function $\Phi(a_2, a_1, a_0)$, they nevertheless throw some light on the general properties inherent in the stress-strain equations.

These equations completely clarify, in particular, the variable elements of the stress-strain relation, i. e., those elements which vary under a change of coordinate systems. Only the three invariant coefficients $\frac{\partial \Phi}{\partial a_2}, \frac{\partial \Phi}{\partial a_1}, \frac{\partial \Phi}{\partial a_0}$ remain indefinite; they are different for different materials and are determined experimentally.

In conclusion, we note that (III.18) and(III.36) imply that equation (III.17) can be rewritten as

$$\delta A = \iiint (\overset{*}{\sigma}_{xx} \delta \varepsilon_{xx} + \overset{*}{\sigma}_{yy} \delta \varepsilon_{yy} + \overset{*}{\sigma}_{zz} \delta \varepsilon_{zz} + \overset{*}{\sigma}_{xy} \delta \varepsilon_{xy} + \overset{*}{\sigma}_{xz} \delta \varepsilon_{xz} + \overset{*}{\sigma}_{yz} \delta \varepsilon_{yz})\, dx dy dz. \qquad \text{(III.45)}$$

This equation is a generalization of the analogous expression of the classical theory of elasticity to the case of deformations of arbitrary magnitude.

§ 29. Boundary Conditions

Equations (III.33) express the conditions which must be satisfied at those points of the bounding surface where the surface loading is prescribed, but the displacements are not.

In virtue of (III.23), (III.33) assumes the form

$$\frac{\partial\Phi}{\partial u_x}\cos(n, X)+\frac{\partial\Phi}{\partial u_y}\cos(n, Y)+\frac{\partial\Phi}{\partial u_z}\cos(n, Z)=\frac{S_n^*}{S_n}f_\xi^*$$

$$\frac{\partial\Phi}{\partial v_x}\cos(n, X)+\frac{\partial\Phi}{\partial v_y}\cos(n, Y)+\frac{\partial\Phi}{\partial v_z}\cos(n, Z)=\frac{S_n^*}{S_n}f_\eta^* \qquad \text{(III.46)}$$

$$\frac{\partial\Phi}{\partial w_x}\cos(n, X)+\frac{\partial\Phi}{\partial w_y}\cos(n, Y)+\frac{\partial\Phi}{\partial w_z}\cos(n, Z)=\frac{S_n^*}{S_n}f_\zeta^*.$$

Here $n\hat{X}$, $n\hat{Y}$ $n\hat{Z}$ are the direction angles relative to the X-, Y-, Z-axes of the normal to the bounding surface before deformation, and f_ξ^*, f_η^*, f_ζ^* are the projections on the same axes of the specific surface force acting at the corresponding point of the surface of the deformed body.

Comparison of equations (III.34) with the three equations obtained by projecting the vector equations (II.39) on the X-, Y-, Z-axes yields

$$\frac{S_x^*}{S_x}\sigma_{n_1,\xi}=\frac{\partial\Phi}{\partial u_x}, \quad \frac{S_y^*}{S_y}\sigma_{n_2,\xi}=\frac{\partial\Phi}{\partial u_y}, \quad \frac{S_z^*}{S_z}\sigma_{n_3,\xi}=\frac{\partial\Phi}{\partial u_z},$$

$$\frac{S_x^*}{S_x}\sigma_{n_1,\eta}=\frac{\partial\Phi}{\partial v_x}, \quad \frac{S_y^*}{S_y}\sigma_{n_2,\eta}=\frac{\partial\Phi}{\partial v_y}, \quad \frac{S_z^*}{S_z}\sigma_{n_3,\eta}=\frac{\partial\Phi}{\partial v_z}, \qquad \text{(III.47)}$$

$$\frac{S_x^*}{S_x}\sigma_{n_1,\zeta}=\frac{\partial\Phi}{\partial w_x}, \quad \frac{S_y^*}{S_y}\sigma_{n_2,\zeta}=\frac{\partial\Phi}{\partial w_y}, \quad \frac{S_z^*}{S_z}\sigma_{n_3,\zeta}=\frac{\partial\Phi}{\partial w_z},$$

where

$\sigma_{n_1,\xi}$, $\sigma_{n_1,\eta}$, $\sigma_{n_1,\zeta}$ are the projections of the stress σ_{n_1} on X, Y, Z,

$\sigma_{n_2,\xi}$, $\sigma_{n_2,\eta}$, $\sigma_{n_2,\zeta}$ are the projections of the stress σ_{n_2} on X, Y, Z,

$\sigma_{n_3,\xi}$, $\sigma_{n_3,\eta}$, $\sigma_{n_3,\zeta}$ are the projections of the stress σ_{n_3} on X, Y, Z.

The equations (III.47) may now be written in the following form:

$$\frac{S_x^*}{S_x}\sigma_{n_1,\xi}\cos(n,X)+\frac{S_y^*}{S_y}\sigma_{n_2,\xi}\cos(n,Y)+\frac{S_z^*}{S_z}\sigma_{n_3,\xi}\cos(n,Z)=\frac{S_n^*}{S_n}f_\xi^*,$$
$$\frac{S_x^*}{S_x}\sigma_{n_1,\eta}\cos(n,X)+\frac{S_y^*}{S_y}\sigma_{n_2,\eta}\cos(n,Y)+\frac{S_z^*}{S_z}\sigma_{n_3,\eta}\cos(n,Z)=\frac{S_n^*}{S_n}f_\eta^*, \qquad \text{(III.48)}$$
$$\frac{S_x^*}{S_x}\sigma_{n_1,\zeta}\cos(n,X)+\frac{S_y^*}{S_y}\sigma_{n_2,\zeta}\cos(n,Y)+\frac{S_z^*}{S_z}\sigma_{n_3,\zeta}\cos(n,Z)=\frac{S_n^*}{S_n}f_\zeta^*,$$

where the projections of the stresses σ_{n_1}, σ_{n_2}, σ_{n_3} can be expressed (with the aid of (II.12) and (III.36)) in terms of the strain components. This enables us to regard the left-hand sides of the expressions (III.48) as functions of the displacements u, v, w.

After these substitutions have been made, the given expressions become the mathematical formulation of the conditions which must be imposed on the displacements at those points of the bounding surface of the body at which u, v, w are not given directly.

§ 30. The Simplification of the Derived Equations in the Case of a Small Deformation

If the deformation is small, its components can be neglected in those equations where they appear together with terms of order unity. This enables us, in the determination of stresses, to identify the dimensions of areas before and after deformation, which is equivalent to identifying the generalized stresses σ_{ij}^* with the true stresses σ_{ij}. The corresponding simplification can be introduced into equations (III.38), (III.43), and (III.48).

If the angles of rotation, as well as the strain components, are small compared to unity, then (as was shown in Ch. II) equations (III.35) can be simplified by neglecting e_{ij} in comparison with unity and with ω_j.

The only other simplification possible consists in neglecting both e_{ij} and also ω_j in (III.35). These equations then assume the form

$$\begin{aligned}
&\frac{\partial}{\partial x}\left[\frac{\partial \Phi}{\partial \varepsilon_{xx}}\right]+\frac{\partial}{\partial y}\left[\frac{\partial \Phi}{\partial \varepsilon_{xy}}\right]+\frac{\partial}{\partial z}\left[\frac{\partial \Phi}{\partial \varepsilon_{xz}}\right]+F_{\xi}=0,\\
&\frac{\partial}{\partial x}\left[\frac{\partial \Phi}{\partial \varepsilon_{xy}}\right]+\frac{\partial}{\partial y}\left[\frac{\partial \Phi}{\partial \varepsilon_{yy}}\right]+\frac{\partial}{\partial z}\left[\frac{\partial \Phi}{\partial \varepsilon_{yz}}\right]+F_{\eta}=0, \quad \text{(III.49)}\\
&\frac{\partial}{\partial x}\left[\frac{\partial \Phi}{\partial \varepsilon_{xz}}\right]+\frac{\partial}{\partial y}\left[\frac{\partial \Phi}{\partial \varepsilon_{yz}}\right]+\frac{\partial}{\partial z}\left[\frac{\partial \Phi}{\partial \varepsilon_{zz}}\right]+F_{\zeta}=0.
\end{aligned}$$

The last simplification is equivalent to assuming that the equations

$$\begin{aligned}
&\frac{\partial \Phi}{\partial \varepsilon_{xx}}=\frac{\partial \Phi}{\partial u_x}, \quad \frac{\partial \Phi}{\partial \varepsilon_{xy}}=\frac{\partial \Phi}{\partial u_y}=\frac{\partial \Phi}{\partial v_x}, \quad \frac{\partial \Phi}{\partial \varepsilon_{xz}}=\frac{\partial \Phi}{\partial u_z}=\frac{\partial \Phi}{\partial u_x},\\
&\frac{\partial \Phi}{\partial \varepsilon_{yy}}=\frac{\partial \Phi}{\partial v_y}, \quad \frac{\partial \Phi}{\partial \varepsilon_{yz}}=\frac{\partial \Phi}{\partial v_z}=\frac{\partial \Phi}{\partial w_y}, \quad \frac{\partial \Phi}{\partial \varepsilon_{zz}}=\frac{\partial \Phi}{\partial w_x} \qquad \text{(III.50)}
\end{aligned}$$

are valid. This corresponds to the determination of the strain components by formulas (I.120), i. e., by the equations of the classical theory.

With this degree of accuracy, the boundary conditions (III.48) can be written in the form

$$\begin{aligned}
&\sigma_{\xi\xi}\cos(n, X)+\sigma_{\xi\eta}\cos(n, Y)+\sigma_{\xi\zeta}\cos(n, Z)=f_{\xi},\\
&\sigma_{\eta\xi}\cos(n, X)+\sigma_{\eta\eta}\cos(n, Y)+\sigma_{\eta\zeta}\cos(n, Z)=f_{\eta} \quad \text{(III.51)}\\
&\sigma_{\zeta\xi}\cos(n, X)+\sigma_{\zeta\eta}\cos(n, Y)+\sigma_{\zeta\zeta}\cos(n, Z)=f_{\zeta},
\end{aligned}$$

where $\sigma_{\xi\xi}, \ldots. \sigma_{\zeta\zeta}$ are the stress components on the areas perpendicular to the X-, Y-, Z-axes.

Using equations (III.51) corresponds to subjecting the solution to boundary conditions without taking the deformation of the body into account. As is well known, this is the procedure in the classical theory.

All the simplifications described above are analogous to those repeatedly examined in Chapters I and II. They can be collectively referred to as "geometric" simplifications, since they are based either on neglecting the changes in certain dimensions or on neglecting the changes in certain angles. We now have to examine the possibilities of simplifying the relations between the stress and strain components, i. e., equations (III.38) or (III. 43) Although these, as well as the previous simplifications, depend on the magnitude of the deformations, the latter must be compared with certain physical constants characteristic of the given material, rather than with unity.

Let us represent the function $\Phi(a_2, a_1, a_0)$ as a power series in the three parameters a_2, a_1, a_0. No negative powers can appear in the series, for otherwise the specific strain energy would tend to infinity for infinitesimal displacements of the points of the body from their initial position, which is absurd.

Furthermore, if we take the strain energy of the body to be zero in the initial state and suppose the body to be free of all stresses in this state, then we must begin the series with terms which contain the strain components to the second power.

Under these conditions, we can write

$$\Phi_3(a_2, a_1, a_0) = A_1 a_2^2 + A_2 a_1 + B_1 a_2^3 + B_2 a_2 a_1 + B_3 a_0$$

$$+C_1a_2^4+C_2a_2^2a_1+C_3a_2a_0+C_4a_1^2$$

$$+D_1a_2^5+D_2a_2^3a_1+D_3a_2^2a_0+D_4a_2a_1^2+D_5a_1a_0$$

$$+\ldots\ldots\ldots\ldots \qquad \text{(III.52)}$$

In this series the A_j's are the coefficients of those terms which contain the strain components to the second power, the B_j's correspond to terms containing the strain components to the third power, the C_j's to those containing the strain components to the fourth power, etc.

The series (III.52) can be regarded as the general expression for the strain energy of an isotropic body which, in its initial state, is free from any internal forces. Its convergence depends both on the magnitudes of the parameters a_2, a_1, a_0 (evidently this improves as they decrease) and on the values of the coefficients $A_j, B_j, C_j, D_j, \ldots$, which are physical constants.

It is therefore impossible to decide a priori at what term the series (III.52) should be terminated, even if it is known that a_2, a_1, a_0 are extremely small compared to unity. This question must be resolved experimentally.

§ 31. Hooke's Law

Let us assume, to begin with, that the strain components are infinitely small.

Then, whatever the relative magnitudes of the physical constants A_j, B_j, C_j, ... (incidentally, they all have the dimension of work/volume; or, if the forces are expressed in kg. and lengths in cm., kg./(cm.)^2), their influence is nullified by the infinite smallness of the strains. Therefore, only those terms in the series (III.52) need be retained which contain the strain components to the smallest (i. e., second) power.

Hence, if the strains are infinitely small, the specific strain energy reduces to

$$\Phi(a_2, a_1, a_0) = A_1 a_2^2 + A_2 a_1. \qquad \text{(III.53)}$$

Substituting this result into (III.38), we arrive at the following stress-strain relation:

$$\begin{aligned} \sigma_{xx} &= 2A_1 a_2 + A_2(\varepsilon_{yy} + \varepsilon_{zz}), & \sigma_{xy} &= \sigma_{yx} = -\frac{1}{2}A_2\varepsilon_{xy}, \\ \sigma_{yy} &= 2A_1 a_2 + A_2(\varepsilon_{xx} + \varepsilon_{zz}), & \sigma_{xz} &= \sigma_{zx} = -\frac{1}{2}A_2\varepsilon_{xz}, \\ \sigma_{zz} &= 2A_1 a_2 + A_2(\varepsilon_{xx} + \varepsilon_{yy}), & \sigma_{yz} &= \sigma_{zy} = -\frac{1}{2}A_2\varepsilon_{yz}. \end{aligned} \qquad \text{(III.54)}$$

If A_1 and A_2 are replaced by two new constants E and μ, where

$$A_2 = -\frac{E}{1+\mu}, \quad 2A_1 + A_2 = \frac{\mu E}{(1+\mu)(1-2\mu)}, \qquad \text{(III.55)}$$

then (III.54) yields

$$\begin{aligned} \sigma_{xx} &= \frac{E}{1+\mu}\left(\varepsilon_{xx} + \frac{\mu}{1-2\mu}a_2\right), & \sigma_{xy} &= \sigma_{yx} = \frac{E}{2(1+\mu)}\varepsilon_{xy}, \\ \sigma_{yy} &= \frac{E}{1+\mu}\left(\varepsilon_{yy} + \frac{\mu}{1-2\mu}a_2\right), & \sigma_{xz} &= \sigma_{zx} = \frac{E}{2(1+\mu)}\varepsilon_{xz}, \\ \sigma_{zz} &= \frac{E}{1+\mu}\left(\varepsilon_{zz} + \frac{\mu}{1-2\mu}a_2\right), & \sigma_{yz} &= \sigma_{zy} = \frac{E}{2(1+\mu)}\varepsilon_{yz}. \end{aligned} \qquad \text{(III.56)}$$

Hence, for infinitely small elongations and shears (but not displacements and angles of rotation, which are in no wise restricted by the preceding assumptions) the stress-strain relation is linear.

Equations (III.56) express the well-known law of Hooke (generalized to spatial deformations). In these equations the constants E and μ correspond to Young's modulus and Poisson's ratio, respectively.

It follows from the above that for every material a range of small deformations can be established for which Hooke's law is approximately valid. It has been shown experimentally that this range differs from material to material.

In the case of steel, the material of greatest interest to technology, the stress-strain relation (Ch. I, § 12 and Fig. 11) is observed to be linear up to elongations of the order $10^{-3} \sim 5\cdot10^{-3}$ (depending on the kind of steel) and up to shears of half this order of magnitude. With further increase of the strains, the character of the stress-strain relation alters radically and a sufficiently precise mathematical description of it in this range requires the retention of a large number of terms in the series (III.52) (even though elongations and shears still remain very small compared to unity).

Hence, as soon as Hooke's law loses its validity, the problem of ascertaining the stress-strain relation is complicated drastically. A further complication arises from the fact that the role of the dissipative forces increases substantially after the limit of proportionality is passed. As a result, only a part of the work done on the body is restored after the load is removed. This renders dubious the application of equations (III.38) to this range of deformations, since they were derived on the assumption that the deformation is completely reversible.

§ 32. On the Applicability of Equations (III.38) to Elastic-Plastic Deformations

Nevertheless, it is not difficult to show that all the basic relations of Hencky's theory of plasticity can be derived from equations (III.38) by introducing suitable assumptions regarding the nature of the dependence of the derivatives $\frac{\partial \Phi}{\partial a_2}, \frac{\partial \Phi}{\partial a_1}, \frac{\partial \Phi}{\partial a_0}$ on the strain components. In order to show this, we shall first of all use equations (III.38) to establish a relation between the two invariants of the stress tensor

$$\begin{gathered} \overset{*}{c}_2 = \overset{*}{\sigma}_{xx} + \overset{*}{\sigma}_{yy} + \overset{*}{\sigma}_{zz} \\ \overset{*}{c}_2^2 - 3\overset{*}{c}_1 = \frac{1}{2}\Big\{(\overset{*}{\sigma}_{xx} - \overset{*}{\sigma}_{yy})^2 + (\overset{*}{\sigma}_{xx} - \overset{*}{\sigma}_{zz})^2 + \\ + (\overset{*}{\sigma}_{yy} - \overset{*}{\sigma}_{zz})^2 + 6\,(\overset{*}{\sigma}_{xy}^2 + \overset{*}{\sigma}_{xz}^2 + \overset{*}{\sigma}_{yz}^2)\Big\} \end{gathered} \tag{III.57}$$

and the strain invariants a_2, a_1, a_0.

Substituting in (III.57) for $\overset{*}{\sigma}_{xx}, \ldots, \overset{*}{\sigma}_{zz}$ their expressions from (III.38) (which is easy to do by referring the latter equations to the principal stresses and strains) we obtain, after some simple calculations,

$$\begin{gathered} \frac{1}{3}\overset{*}{c}_2 = \frac{1}{3}(\overset{*}{\sigma}_{xx} + \overset{*}{\sigma}_{yy} + \overset{*}{\sigma}_{zz}) = \frac{\partial \Phi}{\partial a_2} + \frac{2}{3} a_2 \frac{\partial \Phi}{\partial a_1} \\ + \frac{1}{3} a_1 \frac{\partial \Phi}{\partial a_0}, \\ \overset{*}{c}_2^2 - 3\overset{*}{c}_1 = \left(\frac{\partial \Phi}{\partial a_1}\right)^2 (a_2^2 - 3a_1) + \frac{\partial \Phi}{\partial a_1}\frac{\partial \Phi}{\partial a_0}(a_2 a_1 - 9a_0) \\ + \left(\frac{\partial \Phi}{\partial a_0}\right)^2 (a_1^2 - 3a_2 a_0). \end{gathered} \tag{III.58}$$

Experiments show that the character of only the second of these two relations is drastically changed by the transition from elastic to plastic deformations, while the first (which gives the connection between the average value of the three principal stresses and the strain invariants) changes so little that it can be extended intact, with no serious error, to the plastic range. However, according to (III.56), in the elastic range

$$\frac{1}{3}\left(\overset{*}{\sigma}_{xx} + \overset{*}{\sigma}_{yy} + \overset{*}{\sigma}_{zz}\right) = \frac{1}{3}\frac{E}{1-2\mu} a_2. \tag{III.59}$$

Hence, extending this relation to the plastic range as well, we can write

$$\frac{\partial\Phi}{\partial a_2}+\frac{2}{3}a_2\frac{\partial\Phi}{\partial a_1}+\frac{1}{3}a_1\frac{\partial\Phi}{\partial a_0}=\frac{1}{3}\frac{E}{1-2\mu}a_2. \quad \text{(III.60)}$$

Furthermore, again according to experiment, the stress invariant $c_2^2-3c_1$ can be taken to depend only on the analogous combination of the strain invariants a_2, a_1, i. e., on the quantity $a_2^2-3a_1$. In order to bring the second of equations (III.58) into agreement with this fact it suffices to set $\frac{\partial\Phi}{\partial a_0}$ equal to zero and to regard $\frac{\partial\Phi}{\partial a_1}$ as a function of $a_2^2-3a_1$ alone. Taking into account these assumptions as well as (III.60), equations (III.58) assume the form

$$\begin{gathered}\frac{\partial\Phi}{\partial a_2}-\frac{2}{3}a_2\Psi(T)=\frac{E}{3(1-2\mu)}a_2,\\ S=T\cdot\Psi(T),\end{gathered} \quad \text{(III.61)}$$

where

$$\begin{aligned} S&=\frac{2}{\sqrt{3}}\sqrt{(c_2^2-3c_1)},\\ T&=\frac{2}{\sqrt{3}}\sqrt{(a_2^2-3a_1)},\\ \Psi(T)&=-\frac{\partial\Phi}{\partial a_1}.\end{aligned} \quad \text{(III.62)}$$

In the theory of plasticity the invariant S is usually called the intensity of tangential stresses, and the invariant T the intensity of shearing strain. There is no need to explain the origin of these terms, since a complete exposition of the theory of plasticity is not one of our aims. Returning now to equations (III.38) and substituting in them the values of the derivatives of Φ with respect to the strain invariants obtained as indicated above, we obtain

$$\begin{gathered}\left(\sigma_{xx}-\frac{1}{3}c_2\right)=\Psi(T)\left(\varepsilon_{xx}-\frac{1}{3}a_2\right),\\ \sigma_{xy}=\sigma_{yx}=\frac{1}{2}\Psi(T)\varepsilon_{xy},\end{gathered}$$

$$\left(\sigma_{yy}-\frac{1}{3}c_2\right)=\Psi(T)\left(\varepsilon_{yy}-\frac{1}{3}a_2\right), \qquad \text{(III.63)}$$

$$\sigma_{xz}=\sigma_{zx}=\frac{1}{2}\Psi(T)\,\varepsilon_{xz},$$

$$\left(\sigma_{zz}-\frac{1}{3}c_2\right)=\Psi(T)\left(\varepsilon_{zz}-\frac{1}{3}a_2\right),$$

$$\sigma_{yz}=\sigma_{zy}=\frac{1}{2}\Psi(T)\,\varepsilon_{yz}.$$

These equations are precisely the stress-strain relations proposed by Hencky for elastic-plastic bodies.

Thus the equations of the theory of plasticity are a special case of equations (III.38). In other words, in spite of the irreversibility of a plastic deformation, it can be described by means of equations derived on the explicit assumption that the deformation is reversible. This, at first glance, seems paradoxical. It should, however, be noted that the use of equations (III.38) in the theory of plasticity is admissible only if the process of deformation is an active one, i. e., only if the deformation, during all its intermediate stages, is monotonic in the direction of increasing T. If unloading takes place during deformation, equations (III.63) are no longer valid.

L. M. Kachanov (Kachanov 1, 2) proved, by means of a thermodynamical analysis of an active plastic deformation, that in this case an elastic-plastic body must be indistinguishable in practice from an ideally elastic body with the same stress-strain diagram. This investigation elucidated the physics of the question and showed that there is no paradox in analysing an active plastic deformation with the aid of equations derived for reversible deformations.

§ 33. On the Simplest Variants of Nonlinear Stress-Strain Relations

Let us suppose that the deformations are so large as to render Hooke's law inexact. Then, as a second approximation, one can retain in (III.52) those terms which contain the strain components to the third degree in addition to those containing them to the second degree. Thus, the first five terms in (III.52) are retained, which makes it clear that the description of the elastic properties of the material in this case requires a knowledge of five physical constants. The first two of these are the constants of Hooke's law, since the more general case must reduce to this law in the limit (for infinitely small deformations).

Substituting the expression

$$\Phi(a_2, a_1, a_0) = A_1 a_2^2 + A_2 a_1 + B_1 a_2^3 + B_2 a_2 a_1 + B_3 a_0 \qquad \text{(III.64)}$$

into (III.38), we have

$$\overset{*}{\sigma}_{xx} = (2A_1 + A_2) a_2 + (3B_1 + B_2) a_2^2 + (B_2 + B_3) a_1 - [A_2 + (B_2 + B_3) a_2] \varepsilon_{xx} + B_3 [\varepsilon_{xx}^2 + \tfrac{1}{4} \varepsilon_{xy}^2 + \tfrac{1}{4} \varepsilon_{xz}^2],$$

$$\overset{*}{\sigma}_{yy} = (2A_1 + A_2) a_2 + (3B_1 + B_2) a_2^2 + (B_2 + B_3) a_1 - [A_2 + (B_2 + B_3) a_2] \varepsilon_{yy} + B_3 [\varepsilon_{yy}^2 + \tfrac{1}{4} \varepsilon_{xy}^2 + \tfrac{1}{4} \varepsilon_{yz}^2],$$

$$\overset{*}{\sigma}_{zz} = (2A_1 + A_2) a_2 + (3B_1 + B_2) a_2^2 + (B_2 + B_3) a_1 - [A_2 + (B_2 + B_3) a_2] \varepsilon_{zz} + B_3 [\varepsilon_{zz}^2 + \tfrac{1}{4} \varepsilon_{xz}^2 + \tfrac{1}{4} \varepsilon_{yz}^2],$$

$$\overset{*}{\sigma}_{xy} = -\tfrac{1}{2}\Big\{[A_2 + (B_2 + B_3) a_2] \varepsilon_{xy} - B_3[(\varepsilon_{xx} + \varepsilon_{yy}) \varepsilon_{xy} + \tfrac{1}{2} \varepsilon_{xz} \varepsilon_{yz}]\Big\}, \qquad \text{(III.65)}$$

$$\overset{*}{\sigma}_{xz} = -\frac{1}{2}\Big\{[A_2 + (B_2 + B_3)\,a_2]\,\varepsilon_{xz} - B_3[(\varepsilon_{xx} + \varepsilon_{zz})\,\varepsilon_{xz} + \frac{1}{2}\varepsilon_{xy}\,\varepsilon_{yz}]\Big\},$$

$$\overset{*}{\sigma}_{yz} = -\frac{1}{2}\Big\{[A_2 + (B_2 + B_3)\,a_2]\,\varepsilon_{yz} - B_3[(\varepsilon_{yy} + \varepsilon_{zz})\,\varepsilon_{yz} + \frac{1}{2}\varepsilon_{xy}\,\varepsilon_{xz}]\Big\}.$$

These equations can also be written in the form

$$\overset{*}{\sigma}_{xx} = \frac{E}{1+\mu}\Big\{(1 + \beta_1 a_2)\varepsilon_{xx} + \frac{\mu}{1-2\mu}\,a_2 + \beta_2 a_2^2 - \beta_1 a_1 + \beta_3(\varepsilon_{xx}^2 + \frac{1}{4}\varepsilon_{xy}^2 + \frac{1}{4}\varepsilon_{xz}^2)\Big\},$$

$$\overset{*}{\sigma}_{yy} = \frac{E}{1+\mu}\Big\{(1 + \beta_1 a_2)\varepsilon_{yy} + \frac{\mu}{1-2\mu}\,a_2 + \beta_2 a_2^2 - \beta_1 a_1 + \beta_3(\varepsilon_{yy}^2 + \frac{1}{4}\varepsilon_{xy}^2 + \frac{1}{4}\varepsilon_{yz}^2)\Big\},$$

$$\overset{*}{\sigma}_{zz} = \frac{E}{1+\mu}\Big\{(1 + \beta_1 a_2)\,\varepsilon_{zz} + \frac{\mu}{1-2\mu}\,a_2 + \beta_2 a_2^2 - \beta_1 a_1 + \beta_3(\varepsilon_{zz}^2 + \frac{1}{4}\varepsilon_{xz}^2 + \frac{1}{4}\varepsilon_{yz}^2)\Big\}, \qquad \text{(III.66)}$$

$$\overset{*}{\sigma}_{xy} = \frac{E}{2(1+\mu)}\Big\{(1 + \beta_1 a_2)\,\varepsilon_{xy} + \beta_3[(\varepsilon_{xx} + \varepsilon_{yy})\,\varepsilon_{xy} + \frac{1}{2}\varepsilon_{xz}\,\varepsilon_{yz}]\Big\},$$

$$\overset{*}{\sigma}_{xz} = \frac{E}{2(1+\mu)}\Big\{(1 + \beta_1 a_2)\,\varepsilon_{xz} + \beta_3[(\varepsilon_{xx} + \varepsilon_{zz})\,\varepsilon_{xz} + \frac{1}{2}\varepsilon_{xy}\,\varepsilon_{yz}]\Big\},$$

$$\overset{*}{\sigma}_{yz} = \frac{E}{2(1+\mu)}\Big\{(1 + \beta_1 a_2)\,\varepsilon_{yz} + \beta_3[(\varepsilon_{yy} + \varepsilon_{zz})\,\varepsilon_{yz} + \frac{1}{2}\varepsilon_{xy}\,\varepsilon_{xz}]\Big\},$$

where E, μ, β_1, β_2, β_3 are new constants introduced in place of A_1, A_2, B_1, B_2, and B_3, and are connected with the latter by relations which follow from a comparison of equations (III.65) with equations (III.66).

Comparing (III.66) with Hooke's law (III.56), we

see that the second approximation to the stress-strain law is already much more unwieldy than the first approximation.

It is essential to note that the second approximation differs from the first only in terms which are even functions of the strain components, i. e., terms which remain invariant if the signs of all strain components appearing in them are changed.

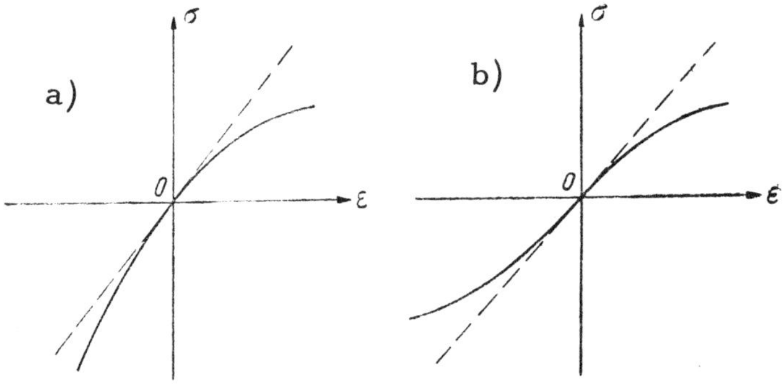

Fig. 16

The extension-compression curve for such a material must lie wholly on one side of its tangent at the origin (Fig. 16a). However, the majority of materials have extension-compression curves of the form shown in Fig. 16b. It follows that deviations from Hooke's law are ordinarily conditioned not so much by terms containing the strains to even powers as by terms containing them to odd powers. In view of this, equations (III.66) by no means yield all possible variants of extension-compression curves.

Nevertheless, the five-constant theory of elasticity, based on the assumption of a specific potential energy of the form (III.64), is at present the subject of considerable attention and has been studied in many papers. The elastic law corresponding to it is ordinarily called Murnaghan's law (Murnaghan 1), although it was actually first proposed much earlier by Voigt (Voigt 1). The first attempt to examine the stress-strain relation in a form different from Hooke's law was made by Bulffinger in a paper (Bulffinger 1) published in the works of the Russian Academy of Sciences in 1729.

In the light of the above remarks, it is not without interest to investigate not only Voigt's five-constant theory of elasticity, but also those forms of the nonlinear stress-strain relation in which the stresses are odd functions of the strains.

To account for such nonlinearity even in the crudest fashion, one must assume the specific strain energy to be of the form

$$\Phi(a_2, a_1, a_0) = A_1 a_2^2 + A_2 a_1 + C_1 a_2^4 + C_2 a_2^2 a_1 + C_3 a_2 a_0 + C_4 a_1^2, \qquad \text{(III.67)}$$

retaining in the series (III.52) all terms containing even powers of the strains up to the fourth inclusive. Then substitution of (III.67) into (III.38) or (III.43), and arguments similar to the preceding, lead to the following stress-strain relations:

$$\overset{*}{\sigma}_{xx} = \frac{E}{1+\mu}\Big\{[1 + \gamma_1 a_2^2 + \gamma_2 a_1]\,\varepsilon_{xx} + \frac{\mu}{1-2\mu} a_2 + \gamma_3 a_2^3 - (2\gamma_1 + \gamma_2 + \gamma_4) a_1 a_2 + \gamma_4 a_0 + \gamma_4 a_2 \Big[\varepsilon_{xx}^2 + \frac{1}{4}\varepsilon_{xy}^2 + \frac{1}{4}\varepsilon_{xz}^2\Big]\Big\},$$

$$\overset{*}{\sigma}_{xy} = \frac{E}{2(1+\mu)}\Big\{[1 + \gamma_1 a_2^2 + \gamma_2 a_1]\,\varepsilon_{xy} + \gamma_4 a_2 [(\varepsilon_{xx} + \varepsilon_{yy})\varepsilon_{xy} + \frac{1}{2}\varepsilon_{xz}\varepsilon_{yz}]\Big\}. \qquad \text{(III.68)}$$

Here E, μ, γ_1, γ_2, γ_3, γ_4 are six physical constants, of which the last five are dimensionless and the first has the dimension of a stress. These equations are even more unwieldy than equations (III.66). Hence, taking account of the nonlinear relation between stresses and strains (even in the crudest approximations) can make the solution of actual problems mathematically unpleasant. This remains true even if we recall that some of the six physical constants which appear in (III.68) can obviously be neglected or can be expressed in terms of one another (this can be established experimentally).

§ 34. Conclusion

It has been shown in this chapter that nonlinearity is introduced into the theory of elasticity in three ways, to wit, through the formulas for the strain components (I.22), through the equations of equilibrium of a volume element of the body (II.43), and through the stress-strain formulas (III.38).

In the first two of the sets of equations mentioned, the retention of the nonlinear terms is conditioned by geometric considerations, i. e., the necessity of taking into account the angles of rotation in determining changes of dimension in the line elements and in formulating the conditions of equilibrium of a volume element.

On the other hand, nonlinear terms appear in the third set of equations if the strain exceeds in magnitude certain physical constants characteristic of the material examined and referred to as the limits of proportionality.

Hence one can speak here of two types of nonlinearity, geometrical and physical. These can be regarded as independent of each other, since, as we repeatedly emphasized in Chapter I, the smallness of the angles of rotation does not imply the smallness of elongations or shears (and conversely).

It follows that there are four types of problems in the theory of elasticity:

1) those having both physical and geometrical linearity;
2) those which are physically nonlinear but geometrically linear;
3) those linear physically but nonlinear geometrically;
4) those nonlinear both physically and geometrically;

In problems of the first type, the angles of rotation are of the same order of magnitude as the elongations and shears, while the elongations do not exceed the limit of proportionality of the given material. The simplest example of this type of problem is the extension of a straight rod by forces which keep the stresses within the limit of proportionality.

In problems of the second type, the angles of rotation can again be neglected in projecting the forces which act on a volume element and in determining strains. However, the elongations exceed the limit of proportionality and this requires a nonlinear stress-strain relation. The example given above becomes a problem of this type if it is complicated by the assumption that the stresses in the rod exceed the limit of proportionality.

In problems of the third type, the angles of rotation are essentially large (with strains not exceeding the limit of proportionality). An example of this type of problem is afforded by the bending of a thin steel strip. It is well known that strips of good steel can straighten out without traces of residual deformation after having their ends brought together. This bears witness to the fact that in these strips, even for large displacements and angles of rotation, the stresses do not exceed the yield point (which, for steel, is close to the limit of proportionality).

Finally, in problems of the fourth type, the strains exceed the limit of proportionality and the angles of rotation are so large that it is necessary to retain nonlinear terms both in the stress-strain formulas and in the equations of equilibrium of an element, as well as in the formulas for the strain components. The preceding example becomes one of this type if it is complicated by assuming that the stresses in the bent strip exceed the limit of proportionality.

It has already been noted that the limit of proportionality for steel is close to the yield point. It follows that the physically nonlinear problems for this material no longer belong to the theory of elasticity but to the theory of plasticity. Hence problems of the second and fourth type will not be examined in the sequel. Nor shall we discuss problems of the first type insofar as they are the subject of the classical (linear) theory of elasticity.

Thus, being concerned with the application to metals of the equations we have derived (and principally to steel) and adhering to the theme "nonlinear theory of *elasticity*", we must in the sequel concentrate on problems of the third type, i. e., those which are physically linear but geometrically nonlinear. In other words, we shall assume in the sequel that the stresses and strains obey Hooke's law (which is equivalent to the assumption that the elongations and shears are very small compared to unity), while the angles of rotation are so large that they cannot be neglected either in determining strains or in the equations of equilibrium of a volume element.

CHAPTER IV

FORMULATION OF ELASTIC PROBLEMS IN TERMS OF STRESSES

§ 35. Two Further Forms for the Equations of Equilibrium of a Volume Element

Returning to the system (II.43), we shall attempt to transform it to a form in which its coefficients are functions of the strain components rather than of the parameters e_{ij}, ω_j.

Remembering that these parameters are given in terms of the displacements u, v, w by equations (I.6), (I.7), and carrying out the differentiation of products in (II.43), we obtain the following three equations

$$\left(1+\frac{\partial u}{\partial x}\right)\left(\frac{\partial \overset{*}{\sigma}_{xx}}{\partial x}+\frac{\partial \overset{*}{\sigma}_{xy}}{\partial y}+\frac{\partial \overset{*}{\sigma}_{xz}}{\partial z}\right)+\frac{\partial u}{\partial y}\left(\frac{\partial \overset{*}{\sigma}_{xy}}{\partial x}\right.$$

$$\left.+\frac{\partial \overset{*}{\sigma}_{yy}}{\partial y}+\frac{\partial \overset{*}{\sigma}_{yz}}{\partial z}\right)+\frac{\partial u}{\partial z}\left(\frac{\partial \overset{*}{\sigma}_{xz}}{\partial x}+\frac{\partial \overset{*}{\sigma}_{yz}}{\partial y}+\frac{\partial \overset{*}{\sigma}_{zz}}{\partial z}\right)+\frac{\partial^2 u}{\partial x^2}\overset{*}{\sigma}_{xx}+\frac{\partial^2 u}{\partial y^2}\overset{*}{\sigma}_{yy}$$

$$+\frac{\partial^2 u}{\partial z^2}\overset{*}{\sigma}_{zz}+2\frac{\partial^2 u}{\partial x \partial y}\overset{*}{\sigma}_{xy}+2\frac{\partial^2 u}{\partial x \partial z}\overset{*}{\sigma}_{xz}+2\frac{\partial^2 u}{\partial z \partial y}\overset{*}{\sigma}_{yz}$$

$$+\frac{V^*}{V}\overset{*}{F}_{\xi}=0, \tag{IV.1}$$

$$\frac{\partial v}{\partial x}\left(\frac{\partial \overset{*}{\sigma}_{xx}}{\partial x}+\frac{\partial \overset{*}{\sigma}_{xy}}{\partial y}+\frac{\partial \overset{*}{\sigma}_{xz}}{\partial z}\right)+\left(1+\frac{\partial v}{\partial y}\right)\left(\frac{\partial \overset{*}{\sigma}_{xy}}{\partial x}+\frac{\partial \overset{*}{\sigma}_{yy}}{\partial y}\right.$$

$$\left.+\frac{\partial \overset{*}{\sigma}_{yz}}{\partial z}\right)+\frac{\partial v}{\partial z}\left(\frac{\partial \overset{*}{\sigma}_{xz}}{\partial x}+\frac{\partial \overset{*}{\sigma}_{yz}}{\partial y}+\frac{\partial \overset{*}{\sigma}_{zz}}{\partial z}\right)+\frac{\partial^2 v}{\partial x^2}\overset{*}{\sigma}_{xx}+\frac{\partial^2 v}{\partial y^2}\overset{*}{\sigma}_{yy}$$

$$+\frac{\partial^2 v}{\partial z^2}\overset{*}{\sigma}_{zz}+2\frac{\partial^2 v}{\partial x \partial y}\overset{*}{\sigma}_{xy}+2\frac{\partial^2 v}{\partial x \partial z}\overset{*}{\sigma}_{xz}+2\frac{\partial^2 v}{\partial y \partial z}\overset{*}{\sigma}_{yz}+\frac{V^*}{V}\overset{*}{F}_{\eta}=0,$$

$$\frac{\partial w}{\partial x}\left(\frac{\partial \overset{*}{\sigma}_{xx}}{\partial x}+\frac{\partial \overset{*}{\sigma}_{xy}}{\partial y}+\frac{\partial \overset{*}{\sigma}_{xz}}{\partial z}\right)+\frac{\partial w}{\partial y}\left(\frac{\partial \overset{*}{\sigma}_{xy}}{\partial x}+\frac{\partial \overset{*}{\sigma}_{yy}}{\partial y}+\frac{\partial \overset{*}{\sigma}_{yz}}{\partial z}\right)$$

$$+\left(1+\frac{\partial w}{\partial z}\right)\left(\frac{\partial \overset{*}{\sigma}_{xz}}{\partial x}+\frac{\partial \overset{*}{\sigma}_{yz}}{\partial y}+\frac{\partial \overset{*}{\sigma}_{zz}}{\partial z}\right)+\frac{\partial^2 w}{\partial x^2}\overset{*}{\sigma}_{xx}+\frac{\partial^2 w}{\partial y^2}\overset{*}{\sigma}_{yy}$$
$$+\frac{\partial^2 w}{\partial z^2}\overset{*}{\sigma}_{zz}+2\frac{\partial^2 w}{\partial x\partial y}\overset{*}{\sigma}_{xy}+2\frac{\partial^2 w}{\partial x\partial z}\overset{*}{\sigma}_{xz}+2\frac{\partial^2 w}{\partial y\partial z}\overset{*}{\sigma}_{yz}+\frac{V^*}{V}F_\zeta^*=0.$$

We multiply the first of these by $1+\frac{\partial u}{\partial x}$, the second by $\frac{\partial v}{\partial x}$, the third by $\frac{\partial w}{\partial x}$, and then add the three resulting equations. Taking into account

$$\begin{aligned}
1+2\varepsilon_{xx}&=\left(1+\frac{\partial u}{\partial x}\right)^2+\left(\frac{\partial v}{\partial x}\right)^2+\left(\frac{\partial w}{\partial x}\right)^2,\\
\varepsilon_{xy}&=\left(1+\frac{\partial u}{\partial x}\right)\frac{\partial u}{\partial y}+\left(1+\frac{\partial v}{\partial y}\right)\frac{\partial v}{\partial x}+\frac{\partial w}{\partial x}\frac{\partial w}{\partial y},\\
\varepsilon_{xz}&=\left(1+\frac{\partial u}{\partial x}\right)\frac{\partial u}{\partial z}+\frac{\partial v}{\partial x}\frac{\partial v}{\partial z}+\left(1+\frac{\partial w}{\partial z}\right)\frac{\partial w}{\partial x},\\
\frac{\partial \varepsilon_{xx}}{\partial x}&=\left(1+\frac{\partial u}{\partial x}\right)\frac{\partial^2 u}{\partial x^2}+\frac{\partial v}{\partial x}\frac{\partial^2 v}{\partial x^2}+\frac{\partial w}{\partial x}\frac{\partial^2 w}{\partial x^2},\\
\frac{\partial \varepsilon_{xy}}{\partial y}-\frac{\partial \varepsilon_{yy}}{\partial x}&=\left(1+\frac{\partial u}{\partial x}\right)\frac{\partial^2 u}{\partial y^2}+\frac{\partial v}{\partial x}\frac{\partial^2 v}{\partial y^2}+\frac{\partial w}{\partial x}\frac{\partial^2 w}{\partial y^2},\\
\frac{\partial \varepsilon_{xz}}{\partial z}-\frac{\partial \varepsilon_{zz}}{\partial x}&=\left(1+\frac{\partial u}{\partial x}\right)\frac{\partial^2 u}{\partial z^2}+\frac{\partial v}{\partial x}\frac{\partial^2 v}{\partial z^2}+\frac{\partial w}{\partial x}\frac{\partial^2 w}{\partial z^2},\\
\frac{\partial \varepsilon_{xx}}{\partial y}&=\left(1+\frac{\partial u}{\partial x}\right)\frac{\partial^2 u}{\partial x\partial y}+\frac{\partial v}{\partial x}\frac{\partial^2 v}{\partial x\partial y}+\frac{\partial w}{\partial x}\frac{\partial^2 w}{\partial x\partial y},\\
\frac{\partial \varepsilon_{xx}}{\partial z}&=\left(1+\frac{\partial u}{\partial x}\right)\frac{\partial^2 u}{\partial x\partial z}+\frac{\partial v}{\partial x}\frac{\partial^2 v}{\partial x\partial z}+\frac{\partial w}{\partial x}\frac{\partial^2 w}{\partial x\partial z},\\
\frac{\partial \varepsilon_{xy}}{\partial z}+\frac{\partial \varepsilon_{xz}}{\partial y}-\frac{\partial \varepsilon_{yz}}{\partial x}&=2\left[\left(1+\frac{\partial u}{\partial x}\right)\frac{\partial^2 u}{\partial y\partial z}+\frac{\partial v}{\partial x}\frac{\partial^2 v}{\partial y\partial z}+\frac{\partial w}{\partial x}\frac{\partial^2 w}{\partial y\partial z}\right],
\end{aligned}\tag{IV.2}$$

the result of this operation yields

$$(1+2\varepsilon_{xx})\left(\frac{\partial \overset{*}{\sigma}_{xx}}{\partial x}+\frac{\partial \overset{*}{\sigma}_{xy}}{\partial y}+\frac{\partial \overset{*}{\sigma}_{xz}}{\partial z}\right)+\varepsilon_{xy}\left(\frac{\partial \overset{*}{\sigma}_{xy}}{\partial x}+\frac{\partial \overset{*}{\sigma}_{yy}}{\partial y}+\frac{\partial \overset{*}{\sigma}_{yz}}{\partial z}\right)$$
$$+\varepsilon_{xz}\left(\frac{\partial \overset{*}{\sigma}_{xz}}{\partial x}+\frac{\partial \overset{*}{\sigma}_{yz}}{\partial y}+\frac{\partial \overset{*}{\sigma}_{zz}}{\partial z}\right)+L_x+\frac{V^*}{V}(1+E_x)F_x^*=0, \tag{IV.3}$$

in which

$$L_x=\overset{*}{\sigma}_{xx}\frac{\partial \varepsilon_{xx}}{\partial x}+\overset{*}{\sigma}_{yy}\left(\frac{\partial \varepsilon_{xy}}{\partial y}-\frac{\partial \varepsilon_{yy}}{\partial x}\right)+\overset{*}{\sigma}_{zz}\left(\frac{\partial \varepsilon_{xz}}{\partial z}-\frac{\partial \varepsilon_{zz}}{\partial x}\right)$$
$$+2\overset{*}{\sigma}_{xy}\frac{\partial \varepsilon_{xx}}{\partial y}+2\overset{*}{\sigma}_{xz}\frac{\partial \varepsilon_{xx}}{\partial z}+\overset{*}{\sigma}_{yz}\left(\frac{\partial \varepsilon_{xy}}{\partial z}+\frac{\partial \varepsilon_{xz}}{\partial y}-\frac{\partial \varepsilon_{yz}}{\partial x}\right), \tag{IV.4}$$

$$F_x^*=\left[\left(1+\frac{\partial u}{\partial x}\right)F_\xi^*+\frac{\partial v}{\partial x}F_\eta^*+\frac{\partial w}{\partial x}F_\zeta^*\right]\frac{1}{1+E_x}. \tag{IV.5}$$

Let us give the physical interpretation of this relation. As seen from Chapter II, the three equations of the system (II.43), and therefore also of the system (IV.1), express the vanishing of the sums of the projections on the X-, Y-, Z-axes of all forces acting on an element of the strained body. On the other hand, in accordance with the table given in § 2,

$$
\begin{aligned}
\cos(X, \mathbf{i}_1) &= \frac{1+e_{xx}}{1+E_x} = \frac{1+\frac{\partial u}{\partial x}}{1+E_x}, \\
\cos(Y, \mathbf{i}_1) &= \frac{\frac{1}{2}e_{xy}+\omega_z}{1+E_x} = \frac{\frac{\partial v}{\partial x}}{1+E_x}, \\
\cos(Z, \mathbf{i}_1) &= \frac{\frac{1}{2}e_{xz}-\omega_y}{1+E_x} = \frac{\frac{\partial w}{\partial x}}{1+E_x}.
\end{aligned}
\tag{IV.6}
$$

Hence it is clear that equations (IV.3) express the fact that the projection on $\mathbf{i}_1$ of the resultant of all the forces acting on an element of the strained body vanishes (Fig. 14).

Also, F_x^* is the projection on the same direction of the specific body force acting at the given point.

Similarly, by adding equations (IV.1) after first multiplying them by $\frac{\partial u}{\partial y}$, $1+\frac{\partial v}{\partial y}$, $\frac{\partial w}{\partial y}$, respectively, and then performing the same operation after multiplying them by $\frac{\partial u}{\partial z}$, $\frac{\partial v}{\partial z}$, $\left(1+\frac{\partial w}{\partial z}\right)$, two more equations analogous to (IV.3) are derived. These express the vanishing of the sums of the projections on the directions $\mathbf{i}_2$ and $\mathbf{i}_3$, of all forces acting on a volume element of the body.

These two equations may be obtained from (IV.3), (IV.4), (IV.5) by a cyclic permutation of the letters x, y, z and a corresponding permutation of u, v, w.

To sum up, we arrive at the following set of differential equations:

$$(1+2\varepsilon_{xx})\left(\frac{\partial\overset{*}{\sigma}_{xx}}{\partial x}+\frac{\partial\overset{*}{\sigma}_{xy}}{\partial y}+\frac{\partial\overset{*}{\sigma}_{xz}}{\partial z}\right)+\varepsilon_{xy}\left(\frac{\partial\overset{*}{\sigma}_{xy}}{\partial x}+\frac{\partial\overset{*}{\sigma}_{yy}}{\partial y}+\frac{\partial\overset{*}{\sigma}_{yz}}{\partial z}\right)$$
$$+\varepsilon_{xz}\left(\frac{\partial\overset{*}{\sigma}_{xz}}{\partial x}+\frac{\partial\overset{*}{\sigma}_{yz}}{\partial y}+\frac{\partial\overset{*}{\sigma}_{zz}}{\partial z}\right)+L_x+\frac{V^*}{V}(1+E_x)\overset{*}{F}_x=0.$$
$$\varepsilon_{xy}\left(\frac{\partial\overset{*}{\sigma}_{xx}}{\partial x}+\frac{\partial\overset{*}{\sigma}_{xy}}{\partial y}+\frac{\partial\overset{*}{\sigma}_{xz}}{\partial z}\right)+(1+2\varepsilon_{yy})\left(\frac{\partial\overset{*}{\sigma}_{xy}}{\partial x}+\frac{\partial\overset{*}{\sigma}_{yy}}{\partial y}+\frac{\partial\overset{*}{\sigma}_{yz}}{\partial z}\right)$$
$$+\varepsilon_{yz}\left(\frac{\partial\overset{*}{\sigma}_{xz}}{\partial x}+\frac{\partial\overset{*}{\sigma}_{yz}}{\partial y}+\frac{\partial\overset{*}{\sigma}_{zz}}{\partial z}\right)+L_y+\frac{V^*}{V}(1+E_y)\overset{*}{F}_y=0. \quad \text{(IV.7)}$$
$$\varepsilon_{xz}\left(\frac{\partial\overset{*}{\sigma}_{xx}}{\partial x}+\frac{\partial\overset{*}{\sigma}_{xy}}{\partial y}+\frac{\partial\overset{*}{\sigma}_{xz}}{\partial z}\right)+\varepsilon_{yz}\left(\frac{\partial\overset{*}{\sigma}_{xy}}{\partial x}+\frac{\partial\overset{*}{\sigma}_{yy}}{\partial y}+\frac{\partial\overset{*}{\sigma}_{yz}}{\partial z}\right)$$
$$+(1+2\varepsilon_{zz})\left(\frac{\partial\overset{*}{\sigma}_{xz}}{\partial x}+\frac{\partial\overset{*}{\sigma}_{yz}}{\partial y}+\frac{\partial\overset{*}{\sigma}_{zz}}{\partial z}\right)+L_z+\frac{V^*}{V}(1+E_z)\overset{*}{F}_z=0,$$

where L_y and L_z are determined from (IV.4) by a cyclic permutation of x, y, z.

Regarding equations (IV.7) as an algebraic system in three unknowns (the expressions in parentheses), and solving this system, we find

$$\frac{\partial\overset{*}{\sigma}_{xx}}{\partial x}+\frac{\partial\overset{*}{\sigma}_{xy}}{\partial y}+\frac{\partial\overset{*}{\sigma}_{xz}}{\partial z}+g^{xx}\left(L_x+\frac{V^*}{V}(1+E_x)\overset{*}{F}_x\right)$$
$$+g^{xy}\left(L_y+\frac{V^*}{V}(1+E_y)\overset{*}{F}_y\right)+g^{xz}\left(L_z+\frac{V^*}{V}(1+E_z)\overset{*}{F}_z\right)=0,$$
$$\frac{\partial\overset{*}{\sigma}_{xy}}{\partial x}+\frac{\partial\overset{*}{\sigma}_{yy}}{\partial y}+\frac{\partial\overset{*}{\sigma}_{yz}}{\partial z}+g^{xy}\left(L_x+\frac{V^*}{V}(1+E_x)\overset{*}{F}_x\right) \quad \text{(IV.8)}$$
$$+g^{yy}\left(L_y+\frac{V^*}{V}(1+E_y)\overset{*}{F}_y\right)+g^{yz}\left(L_z+\frac{V^*}{V}(1+E_z)\overset{*}{F}_z\right)=0,$$
$$\frac{\partial\overset{*}{\sigma}_{xz}}{\partial x}+\frac{\partial\overset{*}{\sigma}_{yz}}{\partial y}+\frac{\partial\overset{*}{\sigma}_{zz}}{\partial z}+g^{xz}\left(L_x+\frac{V^*}{V}(1+E_x)\overset{*}{F}_x\right)$$
$$+g^{yz}\left(L_y+\frac{V^*}{V}(1+E_y)\overset{*}{F}_y\right)+g^{zz}\left(L_z+\frac{V^*}{V}(1+E_z)\overset{*}{F}_z\right)=0,$$

where

$$g^{xx} = \frac{\begin{vmatrix} 1+2\varepsilon_{yy}, & \varepsilon_{yz} \\ \varepsilon_{yz}, & 1+2\varepsilon_{zz} \end{vmatrix}}{D^2} =$$

$$= \frac{1}{D^2}[(1+2\varepsilon_{yy})(1+2\varepsilon_{zz}) - \varepsilon_{yz}^2] = \frac{1}{D^2}\gamma_{xx},$$

$$g^{yy} = \frac{1}{D^2}[(1+2\varepsilon_{xx})(1+2\varepsilon_{zz}) - \varepsilon_{xz}^2] = \frac{1}{D^2}\gamma_{yy}, \qquad \text{(IV.9)}$$

$$g^{zz} = \frac{1}{D^2}[(1+2\varepsilon_{xx})(1+2\varepsilon_{yy}) - \varepsilon_{xy}^2] = \frac{1}{D^2}\gamma_{zz},$$

$$g^{xy} = -\frac{1}{D^2}[\varepsilon_{xy}(1+2\varepsilon_{zz}) - \varepsilon_{yz}\varepsilon_{xz}] = \frac{1}{2D^2}\gamma_{xy},$$

$$g^{xz} = -\frac{1}{D^2}[\varepsilon_{xz}(1+2\varepsilon_{yy}) - \varepsilon_{xy}\varepsilon_{yz}] = \frac{1}{2D^2}\gamma_{xz}, \qquad \text{(IV.10)}$$

$$g^{yz} = -\frac{1}{D^2}[\varepsilon_{yz}(1+2\varepsilon_{xx}) - \varepsilon_{yx}\varepsilon_{zx}] = \frac{1}{2D^2}\gamma_{yz}.$$

Here D is the determinant in equations (I.94) and (I.95), known from Chapter I, and γ_{il} are found in (III.15). The system (IV.8), like (IV.7), has a definite physical meaning which we shall present without proof. Indeed, just as the system (IV.7) expresses the vanishing of the resultant of all forces acting on an element of the body by equating to zero the sums of their *projections* on the directions $\mathbf{i}_1$, $\mathbf{i}_2$, $\mathbf{i}_3$, so the system (IV.8) formulates the same condition by equating to zero the sums of their *components* along the same directions.

Since the vectors $\mathbf{i}_1$, $\mathbf{i}_2$, $\mathbf{i}_3$ are in general not orthogonal, equations (IV.7) and (IV.8) are not identical, even though they state the same physical fact.

We now have three variants of the equations of equilibrium of an element of the strained body:

a) the system (II.43), which expresses the condition for the vanishing of the resultant of all forces acting on an element of the body in terms of the projections of these forces on the fixed X-, Y-, Z-axes;

b) the system (IV.7), which states the same condition by projecting the forces on the directions $\mathbf{i}_1, \mathbf{i}_2, \mathbf{i}_3$;

c) the system (IV.8), which formulates the same condition by resolving these forces in the directions $\mathbf{i}_1, \mathbf{i}_2, \mathbf{i}_3$.

§ 36. Simplification of Equations (IV.7) and (IV.8) for Small Deformations

If the elongations and shears are very small compared to unity, then, as already noted in Chapter II, it is permissible to neglect the deformation of an element of the body in formulating the conditions of equilibrium. In this case, in projecting the forces it is necessary to take into account only the rotation of the element.

In other words, in formulating the equations of statics, we can regard $\mathbf{i}_1, \mathbf{i}_2, \mathbf{i}_3$ as orthogonal. Consequently, we need not distinguish between the projections and components of any force in the directions of these unit vectors. It is clear from the above that then the distinction between (IV.7) and (IV.8) must disappear, and the two equations become identical.

In order to achieve this identity, we must neglect $\varepsilon_{xx}, \varepsilon_{yy}, \varepsilon_{zz}$ as compared to unity in the first system (discarding, moreover, the terms which contain $\varepsilon_{xy}, \varepsilon_{xz}, \varepsilon_{yz}$ as factors) and in the second system we must put

$$\begin{gathered} g^{xx} \approx 1 - 2\varepsilon_{xx} \approx 1, \; g^{xy} \approx -\varepsilon_{xy} \approx 0, \\ g^{xz} \approx -\varepsilon_{xz} \approx 0, \\ g^{yy} \approx 1 - 2\varepsilon_{yy} \approx 1, \; g^{yz} \approx -\varepsilon_{yz} \approx 0, \\ g^{zz} \approx 1 - 2\varepsilon_{zz} \approx 1. \end{gathered} \tag{IV.11}$$

As a result of these approximations, (IV.7) and (IV.8) become

$$
\begin{gathered}
\frac{\partial \overset{*}{\sigma}_{xx}}{\partial x}+\frac{\partial \overset{*}{\sigma}_{xy}}{\partial y}+\frac{\partial \overset{*}{\sigma}_{xz}}{\partial z}+\overset{*}{\sigma}_{yy}\left(\frac{\partial \varepsilon_{xy}}{\partial y}-\frac{\partial \varepsilon_{yy}}{\partial x}\right) \\
+\overset{*}{\sigma}_{zz}\left(\frac{\partial \varepsilon_{xz}}{\partial z}-\frac{\partial \varepsilon_{zz}}{\partial x}\right)+\overset{*}{\sigma}_{xy}\frac{\partial \varepsilon_{xx}}{\partial y}+\overset{*}{\sigma}_{xz}\frac{\partial \varepsilon_{xx}}{\partial z} \\
+\overset{*}{\sigma}_{yz}\left(\frac{\partial \varepsilon_{xy}}{\partial z}+\frac{\partial \varepsilon_{xz}}{\partial y}-\frac{\partial \varepsilon_{yz}}{\partial x}\right)+F_x^*=0,
\end{gathered}
$$

$$
\begin{gathered}
\frac{\partial \overset{*}{\sigma}_{xy}}{\partial x}+\frac{\partial \overset{*}{\sigma}_{yy}}{\partial y}+\frac{\partial \overset{*}{\sigma}_{yz}}{\partial z}+\overset{*}{\sigma}_{xx}\left(\frac{\partial \varepsilon_{xy}}{\partial x}-\frac{\partial \varepsilon_{xx}}{\partial y}\right) \\
+\overset{*}{\sigma}_{zz}\left(\frac{\partial \varepsilon_{yz}}{\partial z}-\frac{\partial \varepsilon_{zz}}{\partial y}\right)+\overset{*}{\sigma}_{xy}\frac{\partial \varepsilon_{yy}}{\partial x} \\
+\overset{*}{\sigma}_{xz}\left(\frac{\partial \varepsilon_{yz}}{\partial x}+\frac{\partial \varepsilon_{xy}}{\partial z}-\frac{\partial \varepsilon_{xz}}{\partial y}\right)+\overset{*}{\sigma}_{yz}\frac{\partial \varepsilon_{yy}}{\partial z}+F_y^*=0,
\end{gathered}
$$

$$
\begin{gathered}
\frac{\partial \overset{*}{\sigma}_{xz}}{\partial x}+\frac{\partial \overset{*}{\sigma}_{yz}}{\partial y}+\frac{\partial \overset{*}{\sigma}_{zz}}{\partial z}+\overset{*}{\sigma}_{xx}\left(\frac{\partial \varepsilon_{xz}}{\partial x}-\frac{\partial \varepsilon_{xx}}{\partial z}\right) \qquad \text{(IV.12)} \\
+\overset{*}{\sigma}_{yy}\left(\frac{\partial \varepsilon_{yz}}{\partial y}-\frac{\partial \varepsilon_{yy}}{\partial z}\right)+\overset{*}{\sigma}_{xy}\left(\frac{\partial \varepsilon_{xz}}{\partial y}+\frac{\partial \varepsilon_{yz}}{\partial x}-\frac{\partial \varepsilon_{xy}}{\partial z}\right) \\
+\overset{*}{\sigma}_{xz}\frac{\partial \varepsilon_{zz}}{dx}+\overset{*}{\sigma}_{yz}\frac{\partial \varepsilon_{zz}}{\partial y}+F_z^*=0.
\end{gathered}
$$

The first of these equations is the condition for the vanishing of the sum of the projections (components) on the direction $\mathbf{i}_1$, of all forces acting on a volume element of the body. Similarly, the other two equations state the same condition for the projections (components) of the forces on the directions $\mathbf{i}_2$ and $\mathbf{i}_3$.

We recall that we are concerned here with an element of the body in the neighborhood of the point x, y, z, which represents, in the unstrained state, a rectangular parallelopiped with edges dx, dy, dz parallel to the X-, Y-, Z-axes.

§ 37. Still Another Form of the Boundary Conditions

We showed in the preceding chapter (§ 29) that equations (III.46) must be satisfied at those points of the bounding surface of the strained body where the forces, rather than the displacements, are prescribed.

If these equations are expanded, they assume the form

$$
\begin{aligned}
&\left[\left(1+\frac{\partial u}{\partial x}\right)\overset{*}{\sigma}_{xx}+\frac{\partial u}{\partial y}\overset{*}{\sigma}_{xy}+\frac{\partial u}{\partial z}\overset{*}{\sigma}_{xz}\right]\cos(n,X)\\
&\quad+\left[\left(1+\frac{\partial u}{\partial x}\right)\overset{*}{\sigma}_{xy}+\frac{\partial u}{\partial y}\overset{*}{\sigma}_{yy}+\frac{\partial u}{\partial z}\overset{*}{\sigma}_{yz}\right]\cos(n,Y)\\
&\quad+\left(1+\frac{\partial u}{\partial x}\right)\overset{*}{\sigma}_{xz}+\frac{\partial u}{\partial y}\overset{*}{\sigma}_{yz}+\frac{\partial u}{\partial z}\overset{*}{\sigma}_{zz}\Big]\cos(n,Z)=\frac{S_n^*}{S_n}f_\xi^*,\\
&\left[\frac{\partial v}{\partial x}\overset{*}{\sigma}_{xx}+\left(1+\frac{\partial v}{\partial y}\right)\overset{*}{\sigma}_{xy}+\frac{\partial v}{\partial z}\overset{*}{\sigma}_{xz}\right]\cos(n,X)\\
&\quad+\left[\frac{\partial v}{\partial x}\overset{*}{\sigma}_{xy}+\left(1+\frac{\partial v}{\partial y}\right)\overset{*}{\sigma}_{yy}+\frac{\partial v}{\partial z}\overset{*}{\sigma}_{yz}\right]\cos(n,Y)\\
&\quad+\left[\frac{\partial v}{\partial x}\overset{*}{\sigma}_{xz}+\left(1+\frac{\partial v}{\partial y}\right)\overset{*}{\sigma}_{yz}+\frac{\partial v}{\partial z}\overset{*}{\sigma}_{zz}\right]\cos(n,Z)=\frac{S_n^*}{S_n}f_\eta^*,\\
&\left[\frac{\partial w}{\partial x}\overset{*}{\sigma}_{xx}+\frac{\partial w}{\partial y}\overset{*}{\sigma}_{xy}+\left(1+\frac{\partial w}{\partial z}\right)\overset{*}{\sigma}_{xz}\right]\cos(n,X)\\
&\quad+\left[\frac{\partial w}{\partial x}\overset{*}{\sigma}_{xy}+\frac{\partial w}{\partial y}\overset{*}{\sigma}_{yy}+\left(1+\frac{\partial w}{\partial z}\right)\overset{*}{\sigma}_{yy}\right]\cos(n,Y)\\
&\quad+\left[\frac{\partial w}{\partial x}\overset{*}{\sigma}_{xz}+\frac{\partial w}{\partial y}\overset{*}{\sigma}_{yz}+\left(1+\frac{\partial w}{\partial z}\right)\overset{*}{\sigma}_{zz}\right]\cos(n,Z)=\frac{S_n^*}{S_n}f_\zeta^*
\end{aligned}
\qquad \text{(IV.13)}
$$

and express the conditions that the projections on the X-, Y-, Z-axes of the stress acting on the bounding surface must be equal to the corresponding projections of the specific surface loading.

However, the fact that the stress vector is equal to the specific loading vector may also be formulated by equating the projections of these two vectors on the directions $\mathbf{i}_1$, $\mathbf{i}_2$, $\mathbf{i}_3$.

To obtain the boundary conditions in this form, we multiply the first of the expressions (IV.13) by $\left(1+\frac{\partial u}{\partial x}\right)$, the second by $\frac{\partial v}{\partial x}$, the third by $\frac{\partial w}{\partial x}$, and then add the resulting equations.

As a result (§ 35), we arrive at the equations

$$\begin{aligned}&[(1+2\varepsilon_{xx})\overset{*}{\sigma}_{xx}+\varepsilon_{xy}\overset{*}{\sigma}_{xy}+\varepsilon_{xz}\overset{*}{\sigma}_{xz}]\cos(n,X)\\&+[(1+2\varepsilon_{xx})\overset{*}{\sigma}_{xy}+\varepsilon_{xy}\overset{*}{\sigma}_{yy}+\varepsilon_{xz}\overset{*}{\sigma}_{yz}]\cos(n,Y)\\&+[(1+2\varepsilon_{xx})\overset{*}{\sigma}_{xz}+\varepsilon_{xy}\overset{*}{\sigma}_{yz}+\varepsilon_{xz}\overset{*}{\sigma}_{zz}]\cos(n,Z)\\&=\frac{S_n^*}{S_n}\left[\left(1+\frac{\partial u}{\partial x}\right)f_\xi^*+\frac{\partial v}{\partial x}f_\eta^*+\frac{\partial w}{\partial x}f_\zeta^*\right].\end{aligned}\tag{IV.14}$$

But, in virtue of (IV.6),

$$\left(1+\frac{\partial u}{\partial x}\right)f_\xi^*+\frac{\partial v}{\partial x}f_\eta^*+\frac{\partial w}{\partial x}f_\zeta^*=(1+E_x)f_x^*,\tag{IV.15}$$

where f_x^* is the projection of the specific surface loading on the direction $\mathbf{i}_1$. Thus,

$$\begin{aligned}&[(1+2\varepsilon_{xx})\overset{*}{\sigma}_{xx}+\varepsilon_{xy}\overset{*}{\sigma}_{xy}+\varepsilon_{xz}\overset{*}{\sigma}_{xz}]\cos(n,X)\\&+[(1+2\varepsilon_{xx})\overset{*}{\sigma}_{xy}+\varepsilon_{xy}\overset{*}{\sigma}_{yy}+\varepsilon_{xz}\overset{*}{\sigma}_{xz}]\cos(n,Y)\\&+[(1+2\varepsilon_{xx})\overset{*}{\sigma}_{xz}+\varepsilon_{xy}\overset{*}{\sigma}_{yz}+\varepsilon_{xz}\overset{*}{\sigma}_{zz}]\cos(n,Z)\\&=\frac{S_n^*}{S_n}(1+E_x)f_x^*.\end{aligned}\tag{IV.16}$$

By a cyclic permutation of the letters x, y, z in the preceding equation, we obtain two more equations, which express the equality of the projections on the directions $\mathbf{i}_2$, $\mathbf{i}_3$, of the stress vector acting on the bounding surface, to the corresponding projections of the specific surface loading. If the elongations and shears are negligibly small compared to unity, we may neglect, in equations (IV.16), all the terms which contain the strain components as factors. As a result, we arrive at the following simplified boundary conditions:

$$\sigma_{xx}\cos(n, X) + \sigma_{xy}\cos(n, Y) + \sigma_{xz}\cos(n, Z) = f_x^*,$$
$$\sigma_{yx}\cos(n, X) + \sigma_{yy}\cos(n, Y) + \sigma_{yz}\cos(n, Z) = f_y^*, \quad \text{(IV.17)}$$
$$\sigma_{zx}\cos(n, X) + \sigma_{zy}\cos(n, Y) + \sigma_{zz}\cos(n, Z) = f_z^*.$$

We must note once again that $\widehat{nX}$, $\widehat{nY}$, $\widehat{nZ}$ are the angles formed by the normal to the bounding surface in the unstrained state, while the σ_{ij} are the components (projections) of the stress in the directions $\mathbf{i}_1$, $\mathbf{i}_2$, $\mathbf{i}_3$, and f_x^*, f_y^*, f_z^* are the projections on the same directions, of the specific surface loading.

§ 38. Simplification of Equations (IV.7) and (IV.8) for Small Angles of Rotation

If the angles of rotation are small compared to unity, but at the same time appreciably exceed the elongations and shears, then (as shown in § 22) the equations of equilibrium can be replaced by the approximations (II.49). The corresponding simplification of (IV.12) is the system

$$\frac{\partial \sigma_{xx}}{\partial x} + \frac{\partial \sigma_{xy}}{\partial y} + \frac{\partial \sigma_{xz}}{\partial z} - \sigma_{yy}\frac{\partial \omega_z}{\partial y} + \sigma_{zz}\frac{\partial \omega_y}{\partial z} - \sigma_{xy}\frac{\partial \omega_z}{\partial x}$$
$$+ \sigma_{xz}\frac{\partial \omega_y}{\partial x} + \sigma_{yz}\left(\frac{\partial \omega_y}{\partial y} - \frac{\partial \omega_z}{\partial z}\right) = -F_x^*,$$
$$\frac{\partial \sigma_{xy}}{\partial x} + \frac{\partial \sigma_{yy}}{\partial y} + \frac{\partial \sigma_{yz}}{\partial z} - \sigma_{zz}\frac{\partial \omega_x}{\partial z} - \sigma_{xx}\frac{\partial \omega_z}{\partial x} - \sigma_{yz}\frac{\partial \omega_x}{\partial y}$$
$$+ \sigma_{xy}\frac{\partial \omega_z}{\partial y} + \sigma_{xz}\left(\frac{\partial \omega_z}{\partial z} - \frac{\partial \omega_x}{\partial x}\right) = -F_y^*, \quad \text{(IV.18)}$$
$$\frac{\partial \sigma_{xz}}{\partial x} + \frac{\partial \sigma_{yz}}{\partial y} + \frac{\partial \sigma_{zz}}{\partial z} - \sigma_{xx}\frac{\partial \omega_y}{\partial x} + \sigma_{yy}\frac{\partial \omega_x}{\partial y} - \sigma_{xz}\frac{\partial \omega_y}{\partial z}$$
$$+ \sigma_{yz}\frac{\partial \omega_x}{\partial z} + \sigma_{xy}\left(\frac{\partial \omega_x}{\partial x} - \frac{\partial \omega_y}{\partial y}\right) = -F_z^*.$$

We then obtain the following approximate equations for angles of rotation which are small in comparison with unity:

$$
\begin{aligned}
\frac{\partial \varepsilon_{yy}}{\partial x} &\approx \frac{\partial \varepsilon_{yy}}{\partial x} - \frac{\partial \varepsilon_{xy}}{\partial y} \approx \frac{\partial \omega_z}{\partial y}, \\
-\frac{\partial \varepsilon_{xx}}{\partial y} &\approx \frac{\partial \varepsilon_{xy}}{\partial x} - \frac{\partial \varepsilon_{xx}}{\partial y} \approx \frac{\partial \omega_z}{\partial x}, \\
\frac{\partial \varepsilon_{xx}}{\partial z} &\approx \frac{\partial \varepsilon_{xx}}{\partial z} - \frac{\partial \varepsilon_{xz}}{\partial x} \approx \frac{\partial \omega_y}{\partial x}, \\
-\frac{\partial \varepsilon_{zz}}{\partial x} &\approx \frac{\partial \varepsilon_{xz}}{\partial z} - \frac{\partial \varepsilon_{zz}}{\partial x} \approx \frac{\partial \omega_y}{\partial z}, \\
\frac{\partial \varepsilon_{zz}}{\partial y} &\approx \frac{\partial \varepsilon_{zz}}{\partial y} - \frac{\partial \varepsilon_{yz}}{\partial z} \approx \frac{\partial \omega_x}{\partial z}, \\
-\frac{\partial \varepsilon_{yy}}{\partial z} &\approx \frac{\partial \varepsilon_{yz}}{\partial y} - \frac{\partial \varepsilon_{yy}}{\partial z} \approx \frac{\partial \omega_x}{\partial y},
\end{aligned}
\qquad \text{(IV.19)}
$$

$$
\begin{aligned}
\frac{\partial \omega_y}{\partial y} - \frac{\partial \omega_z}{\partial z} &\approx \frac{\partial \varepsilon_{xy}}{\partial z} + \frac{\partial \varepsilon_{xz}}{\partial y} - \frac{\partial \varepsilon_{yz}}{\partial x}, \\
\frac{\partial \omega_z}{\partial z} - \frac{\partial \omega_x}{\partial x} &\approx \frac{\partial \varepsilon_{yz}}{\partial x} + \frac{\partial \varepsilon_{xy}}{\partial z} - \frac{\partial \varepsilon_{xz}}{\partial y}, \\
\frac{\partial \omega_x}{\partial x} - \frac{\partial \omega_y}{\partial y} &\approx \frac{\partial \varepsilon_{xz}}{\partial y} + \frac{\partial \varepsilon_{yz}}{\partial x} - \frac{\partial \varepsilon_{xy}}{\partial z}.
\end{aligned}
\qquad \text{(IV.20)}
$$

Equations (IV.19) are consistent only if the derivatives

$$
\frac{\partial \varepsilon_{xy}}{\partial y}, \quad \frac{\partial \varepsilon_{xz}}{\partial x}, \quad \frac{\partial \varepsilon_{yz}}{\partial z}, \quad \frac{\partial \varepsilon_{xy}}{\partial x}, \quad \frac{\partial \varepsilon_{xz}}{\partial z}, \quad \frac{\partial \varepsilon_{yz}}{\partial y} \qquad \text{(IV.21)}
$$

are negligibly small compared to the derivatives

$$
\frac{\partial \varepsilon_{yy}}{\partial x}, \quad \frac{\partial \varepsilon_{xx}}{\partial z}, \quad \frac{\partial \varepsilon_{zz}}{\partial y}, \quad \frac{\partial \varepsilon_{xx}}{\partial y}, \quad \frac{\partial \varepsilon_{zz}}{\partial x}, \quad \frac{\partial \varepsilon_{yy}}{\partial z}, \qquad \text{(IV.21}')
$$

respectively.

There will also be no inconsistency if all (or some) of the derivatives (IV.21′), as well as the corresponding ones in (IV.21), are negligibly small.

It follows from (IV.20) that, in addition to the above condition on the derivatives, it is also necessary to satisfy some one of the following three relations:

$$\frac{\partial\varepsilon_{xy}}{\partial z}+\frac{\partial\varepsilon_{xz}}{\partial y}\approx-\frac{\partial\varepsilon_{yz}}{\partial x},$$
$$\frac{\partial\varepsilon_{yz}}{\partial x}+\frac{\partial\varepsilon_{xy}}{\partial z}\approx-\frac{\partial\varepsilon_{xz}}{\partial y}, \qquad \text{(IV.22)}$$
$$\frac{\partial\varepsilon_{xz}}{\partial y}+\frac{\partial\varepsilon_{y}}{\partial x}\approx-\frac{\partial\varepsilon_{xy}}{\partial z}.$$

Hence the assumption that all the angles of rotation are small compared to unity imposes definite restrictions on the magnitude of the derivatives of the strain components.

Equations (IV.19), (IV.20) establish a relation which must subsist between the derivatives of the strain components and the derivatives of the rotation components in the case under consideration. These equations imply that both sets of derivatives are, in general, of the same order of magnitude. Since we are taking the rotations as appreciably exceeding the elongations and shears, this in turn implies that the strain components must be much more rapidly varying functions of the coordinates than the angles of rotation (at any rate in some directions). In order not to have this derivation contradict the assumption that the deformations are small, it is necessary to assume that the dimensions of the body in the indicated directions are small. This is necessary since, in this case, the growth of the strains, on account of their large gradient, must be limited by the smallness of the range of variation of the corresponding coordinate. This also holds when the angles of rotation are comparable to unity.

Hence, the geometrically nonlinear problems (i. e., problems in which the elongations and shears do not exceed the limits imposed by Hooke's law and the relative rotations of the fibers are large compared to the strains) can arise only in the deformation of bodies having very small dimensions in some directions. Such bodies are thin rods (bodies having small dimensions in the plane of their cross-section), thin plates, and thin shells (bodies having small dimensions in the direction perpendicular to their middle surface).

It is only in the case of such bodies that geometrically nonlinear problems have a real significance.

§ 39. The Generalization of Saint-Venant's Relations to the Case of Large Rotations and Strains

The differential equations derived by Saint-Venant play an essential role in the classical theory of elasticity. They are called either compatability conditions for the strain components or integrability conditions for the equations connecting displacements and strains. With the aid of these relations, problems of the theory of elasticity can be solved directly in terms of stresses without preliminary recourse to displacements. Analogous, but correspondingly more complicated, equations can also be established in the nonlinear theory of elasticity. These equations are identically satisfied when the strain components, expressed in terms of displacements, are substituted into them.

In accordance with the scheme of presentation followed in this book, we shall first obtain these equations without neglecting any quantities (i. e., these equations are then applicable to an arbitrary strain). They will then be simplified to the case in which the elongations and shears are small (but the rotations are arbitrary). The derivation of these equations is very simple if the tensor calculus is used, but becomes very difficult without it. Because of this, the derivation will be carried out in tensor notation, and the reader who may find this difficult is advised to omit the calculations below and to note merely the final result.

We assumed at the very beginning of the book the possibility of introducing a Cartesian coordinate system X, Y, Z. This is equivalent to the assertion that the space considered is Euclidean. If this is the case, the components of the Riemann-Christoffel tensor (the curvature tensor of the space) must all vanish for any curvilinear coordinate system in the space. In particular, they also vanish in the system $\tilde{x}, \tilde{y}, \tilde{z}$ which we adopted as coordinates for the points of a deformed body.

The following expressions hold for the covariant components of the Riemann-Christoffel tensor (see "Vector Calculus and Foundations of Tensor Calculus" by N. E. Kochin, 1938, p. 445. An alternative reference in English is L. P. Eisenhart, An Introduction to Differential Geometry with Use of the Tensor Calculus, Princeton, 1940, p. 103 (Trans.)):

$$R_{i\varkappa\lambda\mu} = \frac{1}{2}\left\{\frac{\partial^2 g_{i\mu}}{\partial x^\varkappa \partial x^\lambda} - \frac{\partial^2 g_{i\lambda}}{\partial x^\varkappa \partial x^\mu} - \frac{\partial^2 g_{\varkappa\mu}}{\partial x^i \partial x^\lambda} + \frac{\partial^2 g_{\varkappa\lambda}}{\partial x^i \partial x^\mu}\right\} - g^{\rho\sigma}\left[\Gamma_{\sigma,\lambda i}\cdot\Gamma_{\rho,\mu\varkappa} - \Gamma_{\sigma,\lambda\varkappa}\cdot\Gamma_{\rho,\mu i}\right], \tag{IV.23}$$

where: g_{jk} are the covariant components of the fundamental tensor in the curvilinear system of coordinates x^1, x^2, x^3,

g^{jk} are the contravariant components of this tensor,

$\Gamma_{l,jk}$ are the Christoffel symbols given in terms of the covariant components of the fundamental tensor by means of the formulas

$$\Gamma_{l,\,jk} = \frac{1}{2}\left(\frac{\partial g_{lj}}{\partial x^k} + \frac{\partial g_{lk}}{\partial x^j} - \frac{\partial g_{jk}}{\partial x^l}\right). \qquad \text{(IV.24)}$$

We note that (according to the customary tensor notation) the symbol for a double sum over the indices σ and ρ is omitted in the right-hand side of formula (IV.23). Let us apply formula (IV.23) to the curvilinear coordinate system $\tilde{x}$, $\tilde{y}$, $\tilde{z}$, setting $\tilde{x}=x^1$, $\tilde{y}=x^2$, $\tilde{z}=x^3$. The covariant components of the fundamental tensor in this coordinate system are equal to

$$\begin{aligned} &g_{xx} = 1 + 2\varepsilon_{xx}, \quad g_{yy} = 1 + 2\varepsilon_{yy}, \quad g_{zz} = 1 + 2\varepsilon_{zz}, \\ &g_{xy} = \varepsilon_{xy}, \quad g_{yx} = \varepsilon_{yx}, \quad g_{yz} = \varepsilon_{yz}. \end{aligned} \qquad \text{(IV.25)}$$

This follows from the expression for the square of the distance between any two points of the deformed body, with the aid of (I.21) and (I.55). The contravariant components of the fundamental tensor (see Kochin, ibid., p. 390, or L. P. Eisenhart, ibid., p. 72 (Trans.)) are determined from (IV.9), (IV.10).

Substituting (IV.9), (IV.25), and (IV.24) into (IV.23), we obtain all the covariant components of the Riemann-Christoffel tensor in the system $\tilde{x}$, $\tilde{y}$, $\tilde{z}$.

In a three-dimensional space only six of these components are distinct (see Kochin, ibid., p. 445, or L. P. Eisenhart, ibid., p. 122, Exercise 4 (Trans.)). Since, in this case, both the g_{kl} and the g^{kl} are expressible in terms of the strain components, equating to zero the components of the Riemann-Christoffel tensor yields six differential equations for the strain components.

Two of these six differential equations are given below. The remaining four can be obtained by cyclic permutation of the letters x, y, z.

$$-\frac{\partial^2\varepsilon_{xx}}{\partial y\partial z}+\frac{1}{2}\frac{\partial}{\partial x}\left(\frac{\partial\varepsilon_{xy}}{\partial z}+\frac{\partial\varepsilon_{xz}}{\partial y}-\frac{\partial\varepsilon_{yz}}{\partial x}\right)$$

$$=g^{xx}\left[\frac{1}{2}\frac{\partial\varepsilon_{xx}}{\partial x}\left(\frac{\partial\varepsilon_{xy}}{\partial z}+\frac{\partial\varepsilon_{xz}}{\partial y}-\frac{\partial\varepsilon_{yz}}{\partial x}\right)-\frac{\partial\varepsilon_{xx}}{\partial y}\frac{\partial\varepsilon_{xx}}{\partial z}\right]$$

$$+g^{yy}\left[\frac{\partial\varepsilon_{yy}}{\partial z}\left(\frac{\partial\varepsilon_{xy}}{\partial x}-\frac{\partial\varepsilon_{xx}}{\partial y}\right)-\frac{1}{2}\frac{\partial\varepsilon_{yy}}{\partial x}\left(\frac{\partial\varepsilon_{xy}}{\partial z}+\frac{\partial\varepsilon_{yz}}{\partial x}-\frac{\partial\varepsilon_{xz}}{\partial y}\right)\right]$$

$$+g^{zz}\left[\frac{\partial\varepsilon_{zz}}{\partial y}\left(\frac{\partial\varepsilon_{xz}}{\partial x}-\frac{\partial\varepsilon_{xx}}{\partial z}\right)-\frac{1}{2}\frac{\partial\varepsilon_{zz}}{\partial x}\left(\frac{\partial\varepsilon_{xz}}{\partial y}+\frac{\partial\varepsilon_{yz}}{\partial x}-\frac{\partial\varepsilon_{xy}}{\partial z}\right)\right]$$

$$+g^{xy}\left[\frac{1}{2}\frac{\partial\varepsilon_{xy}}{\partial x}\left(\frac{\partial\varepsilon_{xy}}{\partial z}+\frac{\partial\varepsilon_{xz}}{\partial y}-\frac{\partial\varepsilon_{yz}}{\partial x}\right)+\frac{\partial\varepsilon_{xx}}{\partial x}\frac{\partial\varepsilon_{yy}}{\partial z}\right.$$

$$\left.-\frac{\partial\varepsilon_{xx}}{\partial y}\frac{\partial\varepsilon_{xy}}{\partial z}-\frac{\partial\varepsilon_{yy}}{\partial x}\frac{\partial\varepsilon_{xx}}{\partial z}\right]+g^{xz}\left[\frac{1}{2}\frac{\partial\varepsilon_{xz}}{\partial x}\left(\frac{\partial\varepsilon_{xy}}{\partial z}\right.\right.$$

$$\left.+\frac{\partial\varepsilon_{xz}}{\partial y}-\frac{\partial\varepsilon_{yz}}{\partial x}\right)+\frac{\partial\varepsilon_{xx}}{\partial x}\frac{\partial\varepsilon_{zz}}{\partial y}-\frac{\partial\varepsilon_{xx}}{\partial z}\frac{\partial\varepsilon_{xz}}{\partial y}$$

$$\left.-\frac{\partial\varepsilon_{xx}}{\partial y}\frac{\partial\varepsilon_{zz}}{\partial x}\right]+g^{yz}\left[\frac{\partial\varepsilon_{yy}}{\partial z}\left(\frac{\partial\varepsilon_{xz}}{\partial x}-\frac{\partial\varepsilon_{xx}}{\partial z}\right)\right.$$

$$+\frac{\partial\varepsilon_{zz}}{\partial y}\left(\frac{\partial\varepsilon_{xy}}{\partial x}-\frac{\partial\varepsilon_{xx}}{\partial y}\right)-\frac{\partial\varepsilon_{yy}}{\partial x}\frac{\partial\varepsilon_{zz}}{\partial x}$$

$$\left.-\frac{1}{4}\left(\frac{\partial\varepsilon_{xz}}{\partial y}+\frac{\partial\varepsilon_{yz}}{\partial x}-\frac{\partial\varepsilon_{xy}}{\partial z}\right)\left(\frac{\partial\varepsilon_{xy}}{\partial z}+\frac{\partial\varepsilon_{yz}}{\partial x}-\frac{\partial\varepsilon_{xz}}{\partial y}\right)\right]; \quad \text{(IV.26)}$$

$$\frac{\partial^2\varepsilon_{xy}}{\partial x\partial y}-\frac{\partial^2\varepsilon_{xx}}{\partial y^2}-\frac{\partial^2\varepsilon_{yy}}{\partial x^2}$$

$$=g^{xx}\left[\frac{\partial\varepsilon_{xx}}{\partial x}\left(\frac{\partial\varepsilon_{xy}}{\partial y}-\frac{\partial\varepsilon_{yy}}{\partial x}\right)-\left(\frac{\partial\varepsilon_{xx}}{\partial y}\right)^2\right]$$

$$+g^{yy}\left[\frac{\partial\varepsilon_{yy}}{\partial y}\left(\frac{\partial\varepsilon_{xy}}{\partial x}-\frac{\partial\varepsilon_{xx}}{\partial y}\right)-\left(\frac{\partial\varepsilon_{yy}}{\partial_x}\right)^2\right]$$

$$+g^{zz}\left[\left(\frac{\partial\varepsilon_{xz}}{\partial x}-\frac{\partial\varepsilon_{xx}}{\partial z}\right)\left(\frac{\partial\varepsilon_{yz}}{\partial y}-\frac{\partial\varepsilon_{yy}}{\partial z}\right)-\frac{1}{4}\left(\frac{\partial\varepsilon_{xz}}{\partial y}\right.\right.$$

$$\left.\left.+\frac{\partial\varepsilon_{yz}}{\partial x}-\frac{\partial\varepsilon_{xy}}{\partial z}\right)^2\right]+g^{xy}\left[\left(\frac{\partial\varepsilon_{xy}}{\partial x}-\frac{\partial\varepsilon_{xx}}{\partial_y}\right)\left(\frac{\partial\varepsilon_{xy}}{\partial y}\right.\right.$$

$$\left.\left.-\frac{\partial\varepsilon_{yy}}{\partial x}\right)+\frac{\partial\varepsilon_{xx}}{\partial x}\frac{\partial\varepsilon_{yy}}{\partial y}-2\frac{\partial\varepsilon_{xx}}{\partial y}\frac{\partial\varepsilon_{yy}}{\partial x}\right]$$

$$+g^{xz}\left[\left(\frac{\partial\varepsilon_{xz}}{\partial x}-\frac{\partial\varepsilon_{xx}}{\partial z}\right)\left(\frac{\partial\varepsilon_{xy}}{\partial y}-\frac{\partial\varepsilon_{yy}}{\partial x}\right)\right.$$

$$\left.+\frac{\partial\varepsilon_{xx}}{\partial x}\left(\frac{\partial\varepsilon_{yz}}{\partial y}-\frac{\partial\varepsilon_{yy}}{\partial z}\right)-\frac{\partial\varepsilon_{xx}}{\partial y}\left(\frac{\partial\varepsilon_{xz}}{\partial y}+\frac{\partial\varepsilon_{yz}}{\partial x}-\frac{\partial\varepsilon_{xy}}{\partial z}\right)\right]$$

$$+g^{yz}\left[\left(\frac{\partial\varepsilon_{xz}}{\partial x}-\frac{\partial\varepsilon_{xx}}{\partial z}\right)\frac{\partial\varepsilon_{yy}}{\partial y}+\left(\frac{\partial\varepsilon_{xy}}{\partial x}-\frac{\partial\varepsilon_{xx}}{\partial y}\right)\left(\frac{\partial\varepsilon_{yz}}{\partial y}\right.\right.$$

$$\left.\left.-\frac{\partial\varepsilon_{yy}}{\partial z}\right)-\frac{\partial\varepsilon_{yy}}{dx}\left(\frac{\partial\varepsilon_{xz}}{\partial y}+\frac{\partial\varepsilon_{yz}}{\partial x}-\frac{\partial\varepsilon_{xy}}{\partial z}\right)\right].$$

The six differential equations obtained remain valid for arbitrary values of the displacements u, v, w. Hence it follows that they are satisfied identically when the strain components are replaced by the displacements in accordance with (I.22).

§ 40. Simplification of the Equations (IV.26) for Small Deformations

If the elongations and shears are very small compared to unity, the difference between the contravariant and covariant components of the Riemann-Christoffel tensor in the curvilinear coordinate system $\tilde{x}, \tilde{y}, \tilde{z}$ is negligible, since it involves only terms of the order of magnitude of the strain components. The terms to be removed from (IV.26) are those which are multiplied by the strain components (but not by their derivatives). This is done in accordance with the simplification adopted in (IV.11).

As a result of this simplification, we obtain the following six approximate equations:

$$\text{I.}\quad \frac{\partial^2 \varepsilon_{xx}}{\partial y^2} + \frac{\partial^2 \varepsilon_{yy}}{\partial x^2} - \frac{\partial^2 \varepsilon_{xy}}{\partial x \partial y} = \left(\frac{\partial \varepsilon_{xx}}{\partial y}\right)^2 + \left(\frac{\partial \varepsilon_{yy}}{\partial x}\right)^2 + \frac{1}{4}\left(\frac{\partial \varepsilon_{yz}}{\partial x} + \frac{\partial \varepsilon_{xz}}{\partial y} - \frac{\partial \varepsilon_{xy}}{\partial z}\right)^2 - \left(\frac{\partial \varepsilon_{xz}}{\partial x} - \frac{\partial \varepsilon_{xx}}{\partial z}\right)\left(\frac{\partial \varepsilon_{yz}}{\partial y} - \frac{\partial \varepsilon_{yy}}{\partial z}\right)$$
$$- \frac{\partial \varepsilon_{yy}}{\partial y}\left(\frac{\partial \varepsilon_{xy}}{\partial x} - \frac{\partial \varepsilon_{xx}}{\partial y}\right) - \frac{\partial \varepsilon_{xx}}{\partial x}\left(\frac{\partial \varepsilon_{xy}}{\partial y} - \frac{\partial \varepsilon_{yy}}{\partial x}\right),$$

$$\text{II.}\quad \frac{\partial^2 \varepsilon_{yy}}{\partial z^2} + \frac{\partial^2 \varepsilon_{zz}}{\partial y^2} - \frac{\partial^2 \varepsilon_{yz}}{\partial y \partial z} = +\left(\frac{\partial \varepsilon_{yy}}{\partial z}\right)^2 + \left(\frac{\partial \varepsilon_{zz}}{\partial y}\right)^2 + \frac{1}{4}\left(\frac{\partial \varepsilon_{xz}}{\partial y} + \frac{\partial \varepsilon_{xy}}{\partial z} - \frac{\partial \varepsilon_{yz}}{\partial x}\right)^2 - \left(\frac{\partial \varepsilon_{yx}}{\partial y} - \frac{\partial \varepsilon_{yy}}{\partial x}\right)\left(\frac{\partial \varepsilon_{xz}}{\partial z} - \frac{\partial \varepsilon_{zz}}{\partial x}\right)$$
$$- \frac{\partial \varepsilon_{zz}}{\partial z}\left(\frac{\partial \varepsilon_{yz}}{\partial y} - \frac{\partial \varepsilon_{yy}}{\partial z}\right) - \frac{\partial \varepsilon_{yy}}{\partial y}\left(\frac{\partial \varepsilon_{yz}}{\partial z} - \frac{\partial \varepsilon_{zz}}{\partial y}\right),$$

$$\text{III.}\quad \frac{\partial^2 \varepsilon_{zz}}{\partial x^2}+\frac{\partial^2 \varepsilon_{xx}}{\partial z^2}-\frac{\partial^2 \varepsilon_{xz}}{\partial x \partial z}=\left(\frac{\partial \varepsilon_{xx}}{\partial z}\right)^2+\left(\frac{\partial \varepsilon_{zz}}{\partial x}\right)^2+\frac{1}{4}\left(\frac{\partial \varepsilon_{xy}}{\partial z}\right.$$
$$\left.+\frac{\partial \varepsilon_{yz}}{\partial x}-\frac{\partial \varepsilon_{xz}}{\partial y}\right)^2-\left(\frac{\partial \varepsilon_{yz}}{\partial z}-\frac{\partial \varepsilon_{zz}}{\partial y}\right)\left(\frac{\partial \varepsilon_{xy}}{\partial x}-\frac{\partial \varepsilon_{xx}}{\partial y}\right)$$
$$-\frac{\partial \varepsilon_{xx}}{\partial x}\left(\frac{\partial \varepsilon_{xz}}{\partial z}-\frac{\partial \varepsilon_{zz}}{\partial x}\right)-\frac{\partial \varepsilon_{zz}}{\partial z}\left(\frac{\partial \varepsilon_{xz}}{\partial x}-\frac{\partial \varepsilon_{xx}}{\partial z}\right),$$

$$\text{IV.}\quad \frac{\partial^2 \varepsilon_{xx}}{\partial y \partial z}-\frac{1}{2}\frac{\partial}{\partial x}\left(\frac{\partial \varepsilon_{xy}}{\partial z}+\frac{\partial \varepsilon_{xz}}{\partial y}-\frac{\partial \varepsilon_{yz}}{\partial x}\right)=\frac{\partial \varepsilon_{xx}}{\partial y}\frac{\partial \varepsilon_{xx}}{\partial z}$$
$$+\frac{1}{2}\frac{\partial \varepsilon_{yy}}{\partial x}\left(\frac{\partial \varepsilon_{xy}}{\partial z}+\frac{\partial \varepsilon_{yz}}{\partial x}-\frac{\partial \varepsilon_{xz}}{\partial y}\right)+\frac{1}{2}\frac{\partial \varepsilon_{zz}}{\partial x}\left(\frac{\partial \varepsilon_{xz}}{\partial y}\right.$$
$$\left.+\frac{\partial \varepsilon_{yz}}{\partial x}-\frac{\partial \varepsilon_{xy}}{\partial z}\right)-\frac{1}{2}\frac{\partial \varepsilon_{xx}}{\partial x}\left(\frac{\partial \varepsilon_{xz}}{\partial y}+\frac{\partial \varepsilon_{xy}}{\partial z}-\frac{\partial \varepsilon_{yz}}{\partial x}\right)$$
$$-\frac{\partial \varepsilon_{yy}}{\partial z}\left(\frac{\partial \varepsilon_{xy}}{\partial x}-\frac{\partial \varepsilon_{xx}}{\partial y}\right)-\frac{\partial \varepsilon_{zz}}{\partial y}\left(\frac{\partial \varepsilon_{xz}}{\partial x}-\frac{\partial \varepsilon_{xx}}{\partial z}\right), \qquad \text{(IV.27)}$$

$$\text{V.}\quad \frac{\partial^2 \varepsilon_{yy}}{\partial x \partial z}-\frac{1}{2}\frac{\partial}{\partial y}\left(\frac{\partial \varepsilon_{yz}}{\partial x}+\frac{\partial \varepsilon_{xy}}{\partial z}-\frac{\partial \varepsilon_{xz}}{\partial y}\right)=\frac{\partial \varepsilon_{yy}}{\partial x}\frac{\partial \varepsilon_{yy}}{\partial z}$$
$$+\frac{1}{2}\frac{\partial \varepsilon_{zz}}{\partial y}\left(\frac{\partial \varepsilon_{yz}}{\partial x}+\frac{\partial \varepsilon_{xz}}{\partial y}-\frac{\partial \varepsilon_{xy}}{\partial z}\right)+\frac{1}{2}\frac{\partial \varepsilon_{xx}}{\partial y}\left(\frac{\partial \varepsilon_{xy}}{\partial z}\right.$$
$$\left.+\frac{\partial \varepsilon_{xz}}{\partial y}-\frac{\partial \varepsilon_{yz}}{\partial x}\right)-\frac{1}{2}\frac{\partial \varepsilon_{yy}}{\partial y}\left(\frac{\partial \varepsilon_{xy}}{\partial z}+\frac{\partial \varepsilon_{yz}}{\partial x}-\frac{\partial \varepsilon_{xz}}{\partial y}\right)$$
$$-\frac{\partial \varepsilon_{zz}}{\partial x}\left(\frac{\partial \varepsilon_{yz}}{\partial y}-\frac{\partial \varepsilon_{yy}}{\partial z}\right)-\frac{\partial \varepsilon_{xx}}{\partial z}\left(\frac{\partial \varepsilon_{xy}}{\partial y}-\frac{\partial \varepsilon_{yy}}{\partial x}\right),$$

$$\text{VI.}\quad \frac{\partial^2 \varepsilon_{zz}}{\partial x \partial y}-\frac{1}{2}\frac{\partial}{\partial z}\left(\frac{\partial \varepsilon_{xz}}{\partial y}+\frac{\partial \varepsilon_{yz}}{\partial x}-\frac{\partial \varepsilon_{xy}}{\partial z}\right)=\frac{\partial \varepsilon_{zz}}{\partial x}\frac{\partial \varepsilon_{zz}}{\partial y}$$
$$+\frac{1}{2}\frac{\partial \varepsilon_{xx}}{\partial z}\left(\frac{\partial \varepsilon_{xz}}{\partial y}+\frac{\partial \varepsilon_{xy}}{\partial z}-\frac{\partial \varepsilon_{yz}}{\partial x}\right)+\frac{1}{2}\frac{\partial \varepsilon_{yy}}{\partial z}\left(\frac{\partial \varepsilon_{yz}}{\partial x}\right.$$
$$\left.+\frac{\partial \varepsilon_{xy}}{\partial z}-\frac{\partial \varepsilon_{xz}}{\partial y}\right)-\frac{1}{2}\frac{\partial \varepsilon_{zz}}{\partial z}\left(\frac{\partial \varepsilon_{yz}}{\partial x}+\frac{\partial \varepsilon_{xz}}{\partial y}-\frac{\partial \varepsilon_{xy}}{\partial z}\right)$$
$$-\frac{\partial \varepsilon_{xx}}{\partial y}\left(\frac{\partial \varepsilon_{xz}}{\partial z}-\frac{\partial \varepsilon_{zz}}{\partial x}\right)-\frac{\partial \varepsilon_{yy}}{\partial x}\left(\frac{\partial \varepsilon_{yz}}{\partial z}-\frac{\partial \varepsilon_{zz}}{\partial y}\right).$$

It is evident that these equations are no longer identically satisfied when the formulas (I.22) are substituted into them. This happens only because of the presence of terms having the strain components as multipliers, i.e., terms which are negligibly small compared to the terms of (IV.27).

Hence equations (IV.27) can be regarded as compatibility conditions for the strain components in the case of small elongations and shears. In this case they may therefore be regarded as generalizations of the Saint-Venant relations.

The presence in (IV.27) of more terms than appear in the Saint-Venant relations is due to the fact that the latter are derived on the assumption that both the angles of rotation and the strains are small compared to unity and are magnitudes of the same order. On the other hand, (IV.27) is derived on the assumption that the strains are small but with no restrictions on the angles of rotation.

§ 41. On the Formulation of the Problems of the Theory of Elasticity in Terms of Stresses and Strains

The results of the first three chapters of this book enable us to write the differential equations of the nonlinear theory of elasticity in terms of displacements and to set up the requisite number of boundary conditions.

The present chapter enables us in principle to solve some problems of the nonlinear theory directly in terms of strains.

Indeed, the equations of equilibrium in the form (IV.7) or (IV.8) do not explicitly contain displacements when the body forces are absent. In that case they contain only stress and strain components. Accordingly, the displacements likewise do not enter explicitly into the boundary conditions (IV.16). Hence, by supplementing equations (IV.7) and (IV.8) with formulas (III.38) and (IV.27), the problems of the nonlinear theory can be solved without first expressing the strains in terms of the displacements. Thus, when body forces are absent and f_x^*, f_y^*, f_z^* are independent of displacements, the problem can be formulated in terms of strains and stresses.

In certain concrete situations, this second procedure may be more convenient than the first (i. e., than solving the problem in terms of displacements). However, the results of the present chapter are of interest even apart from this, because the equations of equilibrium and the boundary conditions, in the form in which they are considered in this chapter, are often convenient even if the problems are treated in terms of displacements.

The relations (IV.27) can be useful in verifying the degree of accuracy of various simplified formulas which are ordinarily applied to determine strains in geometrically nonlinear problems.

We should note in connection with the last remark that, as pointed out in § 38, the geometrically nonlinear theory is applicable to the investigation of the elastic equilibrium of flexible bodies (thin rods, plates, and shells). The problem of the deformation of such bodies can be simplified by means of certain well-established approximations. Thus, for instance, in investigating the plane bending of rods, it can be assumed that the displacements are linear functions of the height of the profile. The latter assumption is equivalent to neglecting the angles of shear compared to the angles of rotation. An analogous method may also be used in the theory of plates and shells. In addition to these simplifications, which affect only the strain equations, the equations of equilibrium are also simplified on the basis of the fact that certain of the stresses arising in the deformation of flexible bodies are always appreciably smaller than the remaining stresses.

All these approximations, which are well-known in the linear theory of plates, shells, and rods, can be completely transferred to the nonlinear theory. These approximations are even more justified in the nonlinear case since the difference between the angles of rotation and the angles of shear is more significant here. Accordingly, the difference between large and small stresses turns out to be more significant. The applicability of the geometrically nonlinear theory to flexible bodies and the resulting simplifications in the formulas greatly facilitate the solution of geometrically nonlinear problems. These approximations amount to a reduction of an arbitrary three-dimensional geometrically nonlinear problem to a two- or even one-dimensional problem. However, in spite of all this, the geometrically nonlinear problems (in view of the nonlinearity of the corresponding differential equations) remain difficult enough. Their exact solution is impossible, as a rule, while an approximate solution leads to complicated calculations.

Nevertheless, a significant number of geometrically nonlinear problems has already been solved and the universal interest which they have aroused leads one to hope that the list of problems solved will continue to increase.

CHAPTER V

THE PROBLEM OF ELASTIC STABILITY

§ 42. Nonuniqueness of Solutions in the Theory of Elasticity

It is known that the differential equations of the linearized theory of elasticity, which are based on Hooke's law and on the omission of nonlinear terms both in the equations for the strain components and in the equations of equilibrium of a volume element, have in each case a unique solution. In other words, the classical theory of elasticity determines a unique position of elastic equilibrium for every body with prescribed load and constraints.

In actuality, however, the solution of such a physical problem is not always unique. The same elastic body, under identical conditions of loading and constraint, may have several possible positions of equilibrium. The incorrect inference to which the classical theory of elasticity leads can be explained by the insufficient accuracy of its formulas, which are derived by neglecting rotations in the expressions for the strain components and in the equations of equilibrium. However, in investigating multiple equilibrium positions of elastic bodies, it is absolutely essential to take the effect of rotations into account.

If there are several possible positions of elastic equilibrium which a body can assume, then, as a rule, not all of them need be stable and hence not all of them are equally likely. Therefore, in considering the problem of elastic equilibrium, we must consider not only the positions of elastic equilibrium but their stability as well. It is essential to observe that when there are several possible positions of equilibrium, that position which is given by the classical theory of elasticity is ordinarily *unstable*. It is clear therefore that the use of the classical theory of elasticity in calculating the strength and deformation of actual structures may lead to catastrophic consequences, since sometimes the actual deformations of the structure differ radically from the predicted ones.

The present chapter is devoted to the theory of stability of the elastic equilibrium of a body of arbitrary shape when it undergoes deformations to which Hooke's law applies. As a point of departure, let us first suppose that the components of the surface and volume forces which act on the body are infinitesimal. Then (for a body of finite rigidity) the displacements of all its points are also infinitesimal. The same is true of the elongations, shears, and rotations. However, an infinitesimal deformation obeys the equations of the classical theory, and hence (in view of the uniqueness theorem mentioned above) every body has just one position of elastic equilibrium for infinitesimal loads, namely the one determined by the classical equations.

Let us suppose, furthermore, that the load on the body is a function of one parameter alone, and let us gradually increase this parameter. The solution remains unique so long as this parameter remains within certain limits. Only when the parameter attains a certain definite value (which we refer to as critical) does there arise the possibility of another position of elastic equilibrium not predicted by the classical theory.

The moment of appearance of a possible bifurcation in the solution corresponds to the critical load. Hence, two positions of equilibrium corresponding to an infinitesimal increment in the critical load differ from one another by an infinitesimal amount.

This last condition will be used as a basis for the determination of critical loads.

The above indicates that if the external load on the body is a function of one parameter alone, then the condition for stable elastic equilibrium is expressed by the inequality

$$p < p_{\text{cr.}},$$

where $p_{\text{cr.}}$ is a function of the shape and dimensions of the body, of its elastic properties, and of the nature of the constraints and loading.

Since, in many problems, the load is indeed characterized by one parameter alone, the above condition is of the greatest practical interest.

Nevertheless, there also arise problems in which the load is a function of two or more parameters $(p, q, r, \ldots)$.

In the latter case, prescribing certain values of the ratios

$$k_1 = q/p, \quad k_2 = r/p, \ldots$$

and then increasing the parameter p, we again arrive at the condition of elastic stability of the structure, in the form of the inequality

$$p < p_{\text{cr.}}.$$

In this case, however, $p_{c.r.}$ is a function not only of the above variables, but also of the ratios $k_1, k_2, \ldots$. It is clear from this that if the load on the body is a function of several parameters, the condition for its elastic stability is expressible in the form

$$f(p, q, r, \ldots) < K,$$

where f is a certain function of the loading parameters and K is a number independent of the load.

§ 43. The Differential Equations which Determine the Critical Loads

Let the load on a body be the critical load. Then there are two possible infinitely close positions of equilibrium.

Denote by u_0, v_0, w_0 the displacements corresponding to that position which becomes unstable when the critical load is reached, and denote the displacements corresponding to the other position by

$$u = u_0 + \alpha u_1, \quad v = v_0 + \alpha v_1, \quad w = w_0 + \alpha w_1. \tag{V.1}$$

Here, $\alpha u\,(x, y, z)$, $\alpha v_1(x, y, z)$, $\alpha w_1(x, y, z)$ are the displacements to which the points of the body must be subjected in order to shift them from the initial position of equilibrium to the new equilibrium position. We shall assume that the functions u_1, v_1, w_1 are finite and that α is an infinitely small quantity independent of x, y, z.

Let us express the strain components which correspond to the second position of equilibrium in terms of the strain components at the initial position and the displacements $\alpha u_1, \alpha v_1, \alpha w_1$. Substituting (V.1) into (I.22) we find

$$\varepsilon_{xx} = \overset{\circ}{\varepsilon}_{xx} + \alpha\varepsilon'_{xx} + \alpha^2\varepsilon''_{xx} \quad \varepsilon_{xy} = \overset{\circ}{\varepsilon}_{xy} + \alpha\varepsilon'_{xy} + \alpha^2\varepsilon''_{xy},$$
$$\varepsilon_{yy} = \overset{\circ}{\varepsilon}_{yy} + \alpha\varepsilon'_{yy} + \alpha^2\varepsilon''_{yy} \quad \varepsilon_{xz} = \overset{\circ}{\varepsilon}_{xz} + \alpha\varepsilon'_{xz} + \alpha^2\varepsilon''_{xz}, \tag{V.2}$$
$$\varepsilon_{zz} = \overset{\circ}{\varepsilon}_{zz} + \alpha\varepsilon'_{zz} + \alpha^2\varepsilon''_{zz} \quad \varepsilon_{yz} = \overset{\circ}{\varepsilon}_{yz} + \alpha\varepsilon'_{yz} + \alpha^2\varepsilon''_{yz},$$

where

$$\overset{\circ}{\varepsilon}_{xx} = \frac{\partial u_0}{\partial x} + \frac{1}{2}\left[\left(\frac{\partial u_0}{\partial x}\right)^2 + \left(\frac{\partial v_0}{\partial x}\right)^2 + \left(\frac{\partial w_0}{\partial x}\right)^2\right],$$
$$\cdots\cdots\cdots\cdots \tag{V.3}$$
$$\overset{\circ}{\varepsilon}_{xy} = \frac{\partial v_0}{\partial x} + \frac{\partial u_0}{\partial y} + \frac{\partial u_0}{\partial x}\frac{\partial u_0}{\partial y} + \frac{\partial v_0}{\partial x}\frac{\partial v_0}{\partial y} + \frac{\partial w_0}{\partial x}\frac{\partial w_0}{\partial y}$$
$$\cdots\cdots\cdots\cdots$$

are the values of the strain components in the first position of equilibrium;

$$\varepsilon'_{xx} = \frac{\partial u_1}{\partial x} + \frac{\partial u_0}{\partial x}\frac{\partial u_1}{\partial x} + \frac{\partial v_0}{\partial x}\frac{\partial v_1}{\partial x} + \frac{\partial w_0}{\partial x}\frac{\partial w_1}{\partial x},$$
$$\cdots\cdots\cdots\cdots \tag{V.4}$$
$$\varepsilon'_{xy} = \frac{\partial v_1}{\partial x} + \frac{\partial u_1}{\partial y} + \frac{\partial u_0}{\partial x}\frac{\partial u_1}{\partial y} + \frac{\partial u_0}{\partial y}\frac{\partial u_1}{\partial x} + \frac{\partial v_0}{\partial x}\frac{\partial v_1}{\partial y} + \frac{\partial v_0}{\partial y}\frac{\partial v_1}{\partial x} + \frac{\partial w_0}{\partial x}\frac{\partial w_1}{\partial y} + \frac{\partial w_0}{\partial y}\frac{\partial w_1}{\partial x}$$
$$\cdots\cdots\cdots\cdots$$

are parameters which depend both on the derivatives of u_0, v_0, w_0 and on the derivatives of u_1, v_1, w_1 ; and

$$\varepsilon''_{xx} = \frac{1}{2}\left[\left(\frac{\partial u_1}{\partial x}\right)^2 + \left(\frac{\partial v_1}{\partial x}\right)^2 + \left(\frac{\partial w_1}{\partial x}\right)^2\right],$$
$$\cdots\cdots\cdots\cdots \tag{V.5}$$
$$\varepsilon''_{xy} = \frac{\partial u_1}{\partial x}\frac{\partial u_1}{\partial y} + \frac{\partial v_1}{\partial x}\frac{\partial v_1}{\partial y} + \frac{\partial w_1}{\partial x}\frac{\partial w_1}{\partial y}$$
$$\cdots\cdots\cdots\cdots$$

are parameters which depend only on the derivatives of u_1, v_1, w_1 and contain only quadratic terms.

Substituting (V.2) into (III.56), i. e., into the formulas expressing Hooke's law, we obtain

$$\sigma_{xx} = \overset{\circ}{\sigma}_{xx} + \alpha\sigma'_{xx} + \alpha^2\sigma''_{xx}, \quad \sigma_{xy} = \overset{\circ}{\sigma}_{xy} + \alpha\sigma'_{xy} + \alpha^2\sigma''_{yx},$$
$$\sigma_{yy} = \overset{\circ}{\sigma}_{yy} + \alpha\sigma'_{yy} + \alpha^2\sigma''_{yy}, \quad \sigma_{xz} = \overset{\circ}{\sigma}_{xz} + \alpha\sigma'_{xz} + \alpha^2\sigma''_{xz}, \tag{V.6}$$
$$\sigma_{zz} = \overset{\circ}{\sigma}_{zz} + \alpha\sigma'_{zz} + \alpha^2\sigma''_{zz}, \quad \sigma_{yz} = \overset{\circ}{\sigma}_{yz} + \alpha\sigma'_{yz} + \alpha^2\sigma''_{yz},$$

where

$$\overset{\circ}{\sigma}_{xx}=\frac{E}{1+\mu}\left(\overset{\circ}{\varepsilon}_{xx}+\frac{\mu}{1-2\mu}\overset{\circ}{a}_2\right),\quad \overset{\circ}{\sigma}_{xy}=\frac{E}{2(1+\mu)}\overset{\circ}{\varepsilon}_{xy},$$
$$\overset{\circ}{\sigma}_{yy}=\frac{E}{1+\mu}\left(\overset{\circ}{\varepsilon}_{yy}+\frac{\mu}{1-2\mu}\overset{\circ}{a}_2\right),\quad \overset{\circ}{\sigma}_{xz}=\frac{E}{2(1+\mu)}\overset{\circ}{\varepsilon}_{xz},\qquad (V.7)$$
$$\overset{\circ}{\sigma}_{zz}=\frac{E}{1+\mu}\left(\overset{\circ}{\varepsilon}_{zz}+\frac{\mu}{1-2\mu}\overset{\circ}{a}_2\right),\quad \overset{\circ}{\sigma}_{yz}=\frac{E}{2(1+\mu)}\overset{\circ}{\varepsilon}_{yz},$$
$$\overset{\circ}{a}_2=\overset{\circ}{\varepsilon}_{xx}+\overset{\circ}{\varepsilon}_{yy}+\overset{\circ}{\varepsilon}_{zz},$$

and σ'_{ij}, σ''_{ij} are expressed in terms of ε'_{ij} and ε''_{ij} in the same manner as $\overset{\circ}{\sigma}_{ij}$ is expressed in terms of $\overset{\circ}{\varepsilon}_{ij}$.

We observe that, since the parameter α is infinitesimal, those terms in (V.2) and (V.6) which contain the square of α are infinitesimals of the second order. Since these terms are of no significance in this and the following sections, they can be neglected in the sequel. Let us now apply the equations of equilibrium of a volume element derived in the second chapter to the second position of equilibrium of the body. These can be taken in the form (II.48) since, in applying Hooke's law, we must suppose the strain components to be negligibly small compared to unity. Introducing the displacements (V.1) and the stresses (V.6) into these equations and omitting those terms which contain the factor α to a degree higher than the first, we obtain

$$\frac{\partial}{\partial x}\Big\{(1+\overset{\circ}{e}_{xx}+\alpha e'_{xx})\overset{\circ}{\sigma}_{xx}+\alpha(1+\overset{\circ}{e}_{xx})\sigma'_{xx}$$
$$+\left[\frac{1}{2}(\overset{\circ}{e}_{xy}+\alpha e'_{xy})-\overset{\circ}{\omega}_z-\alpha\omega'_z\right]\overset{\circ}{\sigma}_{xy}+\alpha\left(\frac{1}{2}\overset{\circ}{e}_{xy}-\overset{\circ}{\omega}_z\right)\sigma'_{xy}$$
$$+\left[\frac{1}{2}(\overset{\circ}{e}_{xz}+\alpha e'_{xz})+\overset{\circ}{\omega}_y+\alpha\omega'_y\right]\overset{\circ}{\sigma}_{xz}+\alpha\left(\frac{1}{2}\overset{\circ}{e}_{xz}+\overset{\circ}{\omega}_y\right)\sigma'_{xz}\Big\}$$
$$+\frac{\partial}{\partial y}\Big\{(1+\overset{\circ}{e}_{xx}+\alpha e'_{xx})\overset{\circ}{\sigma}_{xy}+\alpha(1+\overset{\circ}{e}_{xx})\sigma'_{xy}$$

$$+\left[\frac{1}{2}(\overset{\circ}{e}_{xy}+\alpha e'_{xy})-\overset{\circ}{\omega}_z-\alpha\omega'_z\right]\overset{\circ}{\sigma}_{yy}+\alpha\left(\frac{1}{2}\overset{\circ}{e}_{xy}-\overset{\circ}{\omega}_z\right)\sigma'_{yy}$$

$$+\left[\frac{1}{2}(\overset{\circ}{e}_{xz}+\alpha e'_{xz})+\overset{\circ}{\omega}_y+\alpha\omega'_y\right]\overset{\circ}{\sigma}_{yz}+\alpha\left(\frac{1}{2}\overset{\circ}{e}_{xz}+\overset{\circ}{\omega}_y\right)\sigma'_{yz}\Big\}$$

$$+\frac{\partial}{\partial z}\Big\{(1+\overset{\circ}{e}_{xx}+\alpha e'_{xx})\overset{\circ}{\sigma}_{xz}+\alpha(1+\overset{\circ}{e}_{xx})\sigma'_{xz}+\left[\frac{1}{2}(\overset{\circ}{e}_{xy}+\alpha e'_{xy})\right.$$

$$\left.-\overset{\circ}{\omega}_z-\alpha\omega'_z\right]\overset{\circ}{\sigma}_{yz}+\alpha\left(\frac{1}{2}\overset{\circ}{e}_{xy}-\overset{\circ}{\omega}_z\right)\sigma'_{yz}+\left[\frac{1}{2}(\overset{\circ}{e}_{xz}+\alpha e'_{xz})\right.$$

$$\left.+\overset{\circ}{\omega}_y+\alpha\omega'_y\right]\overset{\circ}{\sigma}_{zz}+\alpha\left(\frac{1}{2}\overset{\circ}{e}_{xz}+\overset{\circ}{\omega}_y\right)\sigma'_{zz}\Big\}+F^*_\xi=0. \qquad \text{(V.8)}$$

The other two equations of this system can be obtained by cyclic permutation of the indices x, y, z.

In equation (V.8), in addition to the notations adopted earlier, $\overset{\circ}{e}_{ij}$ and $\overset{\circ}{\omega}_j$ are the values of e_{ij}, ω_j for $u=u_0, v=v_0, w=w_0$ and e'_{ij}, ω'_j are the values for $u=u_1$, $v=v_1$, $w=w_1$.

Since the displacements u_0, v_0, w_0, as well as u, v, w, correspond to positions of equilibrium of the points of the body, there must exist in addition to (V.8) also equations of the form

$$\frac{\partial}{\partial x}\Big\{(1+\overset{\circ}{e}_{xx})\overset{\circ}{\sigma}_{xx}+\left(\frac{1}{2}\overset{\circ}{e}_{xy}-\overset{\circ}{\omega}_z\right)\overset{\circ}{\sigma}_{xy}+\left(\frac{1}{2}\overset{\circ}{e}_{xz}+\overset{\circ}{\omega}_y\right)\overset{\circ}{\sigma}_{xz}\Big\}$$

$$+\frac{\partial}{\partial y}\Big\{(1+\overset{\circ}{e}_{xx})\overset{\circ}{\sigma}_{xy}+\left(\frac{1}{2}\overset{\circ}{e}_{xy}-\overset{\circ}{\omega}_z\right)\overset{\circ}{\sigma}_{yy}+\left(\frac{1}{2}\overset{\circ}{e}_{xz}+\overset{\circ}{\omega}_y\right)\overset{\circ}{\sigma}_{yz}\Big\}$$

$$+\frac{\partial}{\partial z}\Big\{(1+\overset{\circ}{e}_{xx})\overset{\circ}{\sigma}_{xz}+\left(\frac{1}{2}\overset{\circ}{e}_{xy}-\overset{\circ}{\omega}_z\right)\overset{\circ}{\sigma}_{yz}+\left(\frac{1}{2}\overset{\circ}{e}_{xz}+\overset{\circ}{\omega}_y\right)\overset{\circ}{\sigma}_{zz}\Big\}$$

$$+F^*_{\xi,0}=0, \qquad \text{(V.9)}$$

which are obtained by referring (II.48) to the initial position of equilibrium. Two more equations are obtained by cyclic permutation of x, y, z. We note that the components F_ξ, F_η, F_ζ of the body force are independent of displacements, i. e., they are the same at both positions of equilibrium. This follows from the fact that the only body force which needs to be taken account of in statical problems of the theory of elasticity is the specific weight of the body, whose components along directions independent of the deformation are constant (unless the variation of the volume of the body due to deformation is taken into account). Thus, after subtracting (V.9) from (V.8) and cancelling α, we arrive at the following system of homogeneous differential equations:

$$
\begin{aligned}
&\frac{\partial}{\partial x}\Big\{ e'_{xx}\overset{\circ}{\sigma}_{xx} + (1+\overset{\circ}{e}_{xx})\,\sigma'_{xx} + \Big(\frac{1}{2}e'_{xy} - \omega'_z\Big)\overset{\circ}{\sigma}_{xy} \\
&+\Big(\frac{1}{2}\overset{\circ}{e}_{xy} - \overset{\circ}{\omega}_z\Big)\sigma'_{xy} + \Big(\frac{1}{2}e'_{xz} + \omega'_y\Big)\overset{\circ}{\sigma}_{xz} + \Big(\frac{1}{2}\overset{\circ}{e}_{xz} + \overset{\circ}{\omega}_y\Big)\sigma'_{xz}\Big\} \\
&+\frac{\partial}{\partial y}\Big\{ e'_{xx}\overset{\circ}{\sigma}_{xy} + (1+\overset{\circ}{e}_{xx})\,\sigma'_{xy} + \Big(\frac{1}{2}e'_{xy} - \omega'_z\Big)\overset{\circ}{\sigma}_{yy} \\
&+\Big(\frac{1}{2}\overset{\circ}{e}_{xy} - \overset{\circ}{\omega}_z\Big)\sigma'_{yy} + \Big(\frac{1}{2}e'_{xz} + \omega'_y\Big)\overset{\circ}{\sigma}_{yz} + \Big(\frac{1}{2}\overset{\circ}{e}_{xz} + \overset{\circ}{\omega}_y\Big)\sigma'_{yz}\Big\} \\
&+\frac{\partial}{\partial z}\Big\{ e'_{xx}\overset{\circ}{\sigma}_{xz} + (1+\overset{\circ}{e}_{xx})\,\sigma'_{xz} + \Big(\frac{1}{2}e'_{xy} - \omega'_z\Big)\overset{\circ}{\sigma}_{yz} \\
&+\Big(\frac{1}{2}\overset{\circ}{e}_{xy} - \overset{\circ}{\omega}_z\Big)\sigma'_{yz} + \Big(\frac{1}{2}e'_{xz} + \omega'_y\Big)\overset{\circ}{\sigma}_{zz} \\
&+\Big(\frac{1}{2}\overset{\circ}{e}_{xz} + \overset{\circ}{\omega}_y\Big)\sigma'_{zz}\Big\} = 0. \qquad \text{(V.10)}
\end{aligned}
$$

Two more equations are obtained by cyclic permutation of x, y, z.

It is essential to observe that (V.10) is **linear** in the derivatives of u_1, v_1, w_1 with respect to x, y, z. This follows from the fact that σ'_{ij}, e'_{ij}, ω'_j appear linearly in (V.10), and are themselves (in virtue of (I.6), (I.7), and (III.54)) linear functions of these derivatives. The system of equations (V.9), which corresponds to the initial position of equilibrium, is, on the other hand, nonlinear in the derivatives of u_0, v_0, w_0.

However, in the majority of practical problems, the angles of rotation which correspond to the initial position of equilibrium of the body are either zero or are of the same order as the elongations and shears. In other words, the formulas of the classical theory can ordinarily be applied to the initial position of equilibrium. In these cases, which are of the greatest practical importance, equations (V.9) assume the form

$$\begin{aligned}
&\frac{\partial \overset{\circ}{\sigma}_{xx}}{\partial x}+\frac{\partial \overset{\circ}{\sigma}_{xy}}{\partial y}+\frac{\partial \overset{\circ}{\sigma}_{xz}}{\partial z}+F_x=0,\\
&\frac{\partial \overset{\circ}{\sigma}_{xy}}{\partial x}+\frac{\partial \overset{\circ}{\sigma}_{yy}}{\partial y}+\frac{\partial \overset{\circ}{\sigma}_{yz}}{\partial z}+F_y=0, \qquad \text{(V.11)}\\
&\frac{\partial \overset{\circ}{\sigma}_{xz}}{\partial x}+\frac{\partial \overset{\circ}{\sigma}_{yz}}{\partial y}+\frac{\partial \overset{\circ}{\sigma}_{zz}}{\partial z}+F_z=0,
\end{aligned}$$

where

$$\begin{aligned}
&\overset{\circ}{\sigma}_{xx}=\frac{E}{1+\mu}\left(\overset{\circ}{e}_{xx}+\frac{\mu}{1-2\mu}\overset{\circ}{b}_2\right), \quad \overset{\circ}{\sigma}_{xy}=\frac{E}{2(1+\mu)}\overset{\circ}{e}_{xy},\\
&\overset{\circ}{\sigma}_{yy}=\frac{E}{1+\mu}\left(\overset{\circ}{e}_{yy}+\frac{\mu}{1-2\mu}\overset{\circ}{b}_2\right), \quad \overset{\circ}{\sigma}_{xz}=\frac{E}{2(1+\mu)}\overset{\circ}{e}_{xz},\\
&\overset{\circ}{\sigma}_{zz}=\frac{E}{1+\mu}\left(\overset{\circ}{e}_{zz}+\frac{\mu}{1-2\mu}\overset{\circ}{b}_2\right), \quad \overset{\circ}{\sigma}_{yz}=\frac{E}{2(1+\mu)}\overset{\circ}{e}_{yz}, \qquad \text{(V.12)}\\
&\overset{\circ}{b}_2=\overset{\circ}{e}_{xx}+\overset{\circ}{e}_{yy}+\overset{\circ}{e}_{zz}=\frac{\partial u_0}{\partial x}+\frac{\partial v_0}{\partial y}+\frac{\partial w_0}{\partial z}.
\end{aligned}$$

Solving the linear system (V.11) with prescribed boundary conditions, which we may assume are unchanged by the deformation of the bounding surfaces, we obtain the displacements u_0, v_0, w_0 and the stresses corresponding to the initial position of equilibrium. These quantities become known functions of the coordinates x, y, z and of the still unknown parameter (or parameters) of the external load.

Substituting these functions into (V.10), we arrive at a set of linear homogeneous differential equations for u_1, v_1, w_1.

If homogeneous boundary conditions, to be explained below, are prescribed, the system has nontrivial (i. e., nonidentically vanishing) solutions only for certain definite values of the load parameter. These are the characteristic values of the system. To each such characteristic value there corresponds a point of bifurcation of the solution of the equations.

This is a method for investigating elastic stability.

It is to be noted that the assumption that the classical theory is applicable to the initial position of equilibrium implies that equations (V.10) can also be simplified. Indeed, in passing from (V.9) to (V.11), it was necessary to neglect all terms in the first system which have $\overset{\circ}{e}_{ij}$, $\overset{\circ}{\omega}_j$ as coefficients. As is known (§ 23), this is equivalent to neglecting the angles of rotation of the line elements of the body in projecting the forces which act on a volume element of the body.

This being the case, however, there are no grounds for retaining the analogous terms in (V.10).

Hence this system can be written in the following form: (V.13)

$$\frac{\partial}{\partial x}\left\{\sigma'_{xx}+e'_{xx}\overset{\circ}{\sigma}_{xx}+\left(\frac{1}{2}e'_{xy}-\omega'_z\right)\overset{\circ}{\sigma}_{xy}+\left(\frac{1}{2}e'_{xz}+\omega'_y\right)\overset{\circ}{\sigma}_{xz}\right\}$$
$$+\frac{\partial}{\partial y}\left\{\sigma'_{xy}+e'_{xx}\overset{\circ}{\sigma}_{xy}+\left(\frac{1}{2}e'_{xy}-\omega'_z\right)\overset{\circ}{\sigma}_{yy}+\left(\frac{1}{2}e'_{xz}+\omega'_y\right)\overset{\circ}{\sigma}_{yz}\right\}$$
$$+\frac{\partial}{\partial z}\left\{\sigma'_{xz}+e'_{xx}\overset{\circ}{\sigma}_{xz}+\left(\frac{1}{2}e'_{xy}-\omega'_z\right)\overset{\circ}{\sigma}_{yz}+\left(\frac{1}{2}e'_{xz}+\omega'_y\right)\overset{\circ}{\sigma}_{zz}\right\}=0.$$

The remaining equations can be obtained by cyclic permutation of x, y, z.

Furthermore, let us examine the expressions (V.4) which we shall first rewrite in the form

$$\varepsilon_{xx}=(1+\overset{\circ}{e}_{xx})\frac{\partial u_1}{\partial x}+\left(\frac{1}{2}\overset{\circ}{e}_{xy}+\overset{\circ}{\omega}_z\right)\frac{\partial v_1}{\partial x}+\left(\frac{1}{2}\overset{\circ}{e}_{xz}-\overset{\circ}{\omega}_y\right)\frac{\partial w_1}{\partial x},$$

. . . .

$$\begin{aligned}\varepsilon'_{xy} &= \frac{\partial u_1}{\partial y} + \frac{\partial v_1}{\partial x} + \overset{\circ}{e}_{xx}\frac{\partial u_1}{\partial y} + \overset{\circ}{e}_{yy}\frac{\partial v_1}{\partial x} + \left(\frac{1}{2}\overset{\circ}{e}_{xy} - \overset{\circ}{\omega}_z\right)\frac{\partial u_1}{\partial x} \\ &+ \left(\frac{1}{2}\overset{\circ}{e}_{xy} + \overset{\circ}{\omega}_z\right)\frac{\partial v_1}{\partial y} + \left(\frac{1}{2}\overset{\circ}{e}_{yz} + \overset{\circ}{\omega}_x\right)\frac{\partial w_1}{\partial x} \\ &+ \left(\frac{1}{2}\overset{\circ}{e}_{xz} - \overset{\circ}{\omega}_y\right)\frac{\partial w_1}{\partial y},\end{aligned} \qquad (V.14)$$

. . . .

In accordance with the approximations noted above, all terms multiplied by $\overset{\circ}{e}_{ij}, \overset{\circ}{\omega}_j$ must be neglected here. We then obtain for the parameters ε'_{ij} (in place of (V.4)) the very much simpler expressions

$$\begin{aligned} \varepsilon'_{xx} &= e'_{xx} = \frac{\partial u_1}{\partial x}, & \varepsilon'_{xy} &= e'_{xy} = \frac{\partial u_1}{\partial y} + \frac{\partial v_1}{\partial x}, \\ \varepsilon'_{yy} &= e'_{yy} = \frac{\partial v_1}{\partial y}, & \varepsilon'_{xz} &= e'_{xz} = \frac{\partial u_1}{\partial z} + \frac{\partial w_1}{\partial x}, \\ \varepsilon'_{zz} &= e'_{zz} = \frac{\partial w_1}{\partial z}, & \varepsilon'_{yz} &= e'_{yz} = \frac{\partial v_1}{\partial z} + \frac{\partial w_1}{\partial y}. \end{aligned} \qquad (V.15)$$

Returning to equations (V.13), we note that the parameters ω'_j occurring in them are proportional to the increments of the angles of rotation which a volume element of the body acquires in passing from the first position of equilibrium to the second. On the other hand, the quantities e'_{ij}, as just shown, are proportional (with the same coefficient of proportionality α) to the increments of the strain components.

A characteristic feature of problems dealing with the loss of stability of elastic equilibrium is the change from positions of equilibrium with small angles of rotation to positions of equilibrium with angles of rotation which substantially exceed the strain components. This enables us to neglect all terms in equations (V.13) which are multiplied by $e'_{ij} = \varepsilon'_{ij}$ as compared with terms that are multiplied by ω'_j. Thus, (V.13) becomes

$$
\begin{gathered}
\frac{\partial}{\partial x}\left[\sigma'_{xx} - \omega'_z \overset{\circ}{\sigma}_{xy} + \omega'_y \overset{\circ}{\sigma}_{xz}\right]
+ \frac{\partial}{\partial y}\left[\sigma'_{xy} - \omega'_z \overset{\circ}{\sigma}_{yy} + \omega'_y \overset{\circ}{\sigma}_{yz}\right]
+ \frac{\partial}{\partial z}\left[\sigma'_{xz} - \omega'_z \overset{\circ}{\sigma}_{yz} + \omega'_y \overset{\circ}{\sigma}_{zz}\right] = 0, \\
\frac{\partial}{\partial x}\left[\sigma'_{xy} - \omega'_x \overset{\circ}{\sigma}_{xz} + \omega'_z \overset{\circ}{\sigma}_{xx}\right]
+ \frac{\partial}{\partial y}\left[\sigma'_{yy} - \omega'_x \overset{\circ}{\sigma}_{yz} + \omega'_z \overset{\circ}{\sigma}_{yx}\right]
+ \frac{\partial}{\partial z}\left[\sigma'_{zy} - \omega'_x \overset{\circ}{\sigma}_{zz} + \omega'_z \overset{\circ}{\sigma}_{zx}\right] = 0, \qquad (V.16) \\
\frac{\partial}{\partial x}\left[\sigma'_{xz} - \omega'_y \overset{\circ}{\sigma}_{xx} + \omega'_x \overset{\circ}{\sigma}_{xy}\right]
+ \frac{\partial}{\partial y}\left[\sigma'_{yz} - \omega'_y \overset{\circ}{\sigma}_{yx} + \omega'_x \overset{\circ}{\sigma}_{yy}\right]
+ \frac{\partial}{\partial z}\left[\sigma'_{zz} - \omega'_y \overset{\circ}{\sigma}_{zx} + \omega'_x \overset{\circ}{\sigma}_{zy}\right] = 0,
\end{gathered}
$$

which we shall regard as final.

Substitution of the expressions for σ'_{ij}, ω'_j in equations (V.16) transforms them into a system of three linear homogeneous differential equations with the three unknowns u_1, v_1, w_1. The determination of critical loads is thus reduced to the determination of the characteristic values of the system (for prescribed boundary conditions) if the classical theory is applicable to the initial state.

§ 44. Boundary Conditions of the Problem of Elastic Stability

Let us formulate the boundary conditions for the system (V.16) using the results of the previous chapter.

They may be obtained by comparing the boundary conditions for the first and second positions of equilibrium.

Assume that geometric boundary conditions are

prescribed on some part of the bounding surface, i. e., that there the displacements are prescribed explicitly. Suppose further that on the remainder of the surface the components f_x^*, f_y^*, f_z^* of the external load are prescribed. The geometric boundary conditions can be written in the form

$$u=\varphi_1(x, y, z),\quad v=\varphi_2(x, y, z),\quad w=\varphi_3(x, y, z), \qquad \text{(V.17)}$$

where $\varphi_1(x, y, z)$, $\varphi_2(x, y, z)$, $\varphi_3(x, y, z)$ are prescribed functions.

These equations must be satisfied at both the first and second positions of equilibrium. Hence (on the corresponding sections of the bounding surface), we have

$$\begin{gathered} u_0=\varphi_1,\quad v_0=\varphi_2,\quad w_0=\varphi_3, \\ u_0+\alpha u_1=\varphi_1,\quad v_0+\alpha v_1=\varphi_2,\quad w_0+\alpha w_1=\varphi_3. \end{gathered} \qquad \text{(V.18)}$$

Subtracting the upper equations from the lower, we arrive at the following boundary conditions for u_1, v_1, w_1:

$$u_1=0,\quad v_1=0,\quad w_1=0, \qquad \text{(V.19)}$$

which must be satisfied at those boundary points of the body at which the displacements are explicitly prescribed. We note that, at the points of the body where the displacements are prescribed, u_1, v_1, w_1 must satisfy the same conditions as were imposed on the variations of the displacements. This fact will be used in the next section. Let us, further, write the boundary conditions for the first and second positions of equilibrium at those points at which the external surface load is prescribed.

If it is assumed that the linear theory of elasticity applies to the initial position of equilibrium, the conditions for the equality of forces on the surface of the body assume the form

$$
\begin{aligned}
&\overset{\circ}{\sigma}_{xx}\cos(n,X)+\overset{\circ}{\sigma}_{xy}\cos(n,Y)+\overset{\circ}{\sigma}_{xz}\cos(n,Z)\\
&\qquad=f_\xi(u_0,v_0,w_0)=\overset{\circ}{f}_\xi,\\
&\overset{\circ}{\sigma}_{xy}\cos(n,X)+\overset{\circ}{\sigma}_{yy}\cos(n,Y)+\overset{\circ}{\sigma}_{yz}\cos(n,Z)\\
&\qquad=f_\eta(u_0,v_0,w_0)=\overset{\circ}{f}_\eta,\\
&\overset{\circ}{\sigma}_{xz}\cos(n,X)+\overset{\circ}{\sigma}_{yz}\cos(n,Y)+\overset{\circ}{\sigma}_{zz}\cos(n,Z)\\
&\qquad=f_\zeta(u_0,v_0,w_0)=\overset{\circ}{f}_\zeta,
\end{aligned}
\tag{V.20}
$$

where $f_\xi(u_0, v_0, w_0)$, $f_\eta(u_0, v_0, w_0)$, $f_\zeta(u_0, v_0, w_0)$ are the projections of the specific surface load on the X-, Y-, Z-axes at the first position of equilibrium. The analogous conditions for the second position of equilibrium (adjacent to the first) are

$$
\begin{aligned}
&\sigma_{n_1,\xi}\cos(n,X)+\sigma_{n_2,\xi}\cos(n,Y)+\sigma_{n_3,\xi}\cos(n,Z)\\
&\qquad=f_\xi(u_0+\alpha u_1,\ v_0+\alpha v_1,\ w_0+\alpha w_0),\\
&\sigma_{n_1,\eta}\cos(n,X)+\sigma_{n_2,\eta}\cos(n,Y)+\sigma_{n_3,\zeta}\cos(n,Z)\\
&\qquad=f_\eta(u_0+\alpha u_1,\ v_0+\alpha v_1,\ w_0+\alpha w_1),\\
&\sigma_{n_1,\zeta}\cos(n,X)+\sigma_{n_2,\zeta}\cos(n,Y)+\sigma_{n_3,\zeta}\cos(n,Z)\\
&\qquad=f_\zeta(u_0+\alpha u_1,\ v_0+\alpha v_1,\ w_0+\alpha w_1),
\end{aligned}
\tag{V.21}
$$

or, noting formulas (II.39), (II.42), and (V.6), as well as the approximations adopted in the preceding section, we obtain

$$
\begin{aligned}
&[\overset{\circ}{\sigma}_{xx}+\alpha(\sigma'_{xx}-\omega'_z\overset{\circ}{\sigma}_{xy}+\omega'_y\overset{\circ}{\sigma}_{xz})]\cos(nX)\\
&\quad+[\overset{\circ}{\sigma}_{xy}+\alpha(\sigma'_{xy}-\omega'_z\overset{\circ}{\sigma}_{yy}+\omega'_y\overset{\circ}{\sigma}_{yz})]\cos(nY)\\
&\quad+[\overset{\circ}{\sigma}_{xz}+\alpha(\sigma'_{xz}-\omega'_z\overset{\circ}{\sigma}_{yz}+\omega'_y\overset{\circ}{\sigma}_{zz})]\cos(nZ)\\
&\quad=f_\xi(u_0+\alpha u_1,\ v_0+\alpha v_1,\ w_0+\alpha w_1).
\end{aligned}
\tag{V.22}
$$

Subtracting formulas (V.20) from formulas of the form (V.22), we arrive at the following formulation of the boundary conditions which must be imposed on the functions u_1, v_1, w_1 at those boundary points of the body at which the external forces are prescribed:

$$\begin{aligned}&(\sigma'_{xx} - \omega'_z \mathring{\sigma}_{xy} + \omega'_y \mathring{\sigma}_{xz}) \cos (nX)\\&\quad + (\sigma'_{xy} - \omega'_z \mathring{\sigma}_{yy} + \omega'_y \mathring{\sigma}_{yz}) \cos (nY)\\&\quad + (\sigma'_{xz} - \omega'_z \mathring{\sigma}_{yz} + \omega'_y \mathring{\sigma}_{zz}) \cos (nZ)\\&= \lim_{\alpha \to 0} \left\{ \frac{1}{\alpha} [f_\xi (u_0 + \alpha u_1, v_0 + \alpha v_1, w_0 + \alpha w_1) - \mathring{f}_\xi] \right\}.\end{aligned} \tag{V.23}$$

Two other boundary conditions are obtainable by cyclic permutation of x, y, z.

Note that, while in the preceding section the body forces can be regarded as independent of the displacements, such an assumption is, in general, impossible for the surface forces. In particular, if the surface of the body is subjected to hydrostatic pressure (the pressure of a liquid or gas), the magnitude of the specific surface load remains invariant under deformation, but its direction changes (since hydrostatic pressure is always directed along the normal to the surface on which it acts). Hence the components f_ξ, f_η, f_ζ of the specific surface load will vary under deformation, with the result that the right-hand sides of formulas of the form (V.23) can no longer be set equal to zero.

For bodies whose surfaces are subjected to hydrostatic pressure, we have

$$\begin{aligned}f_x(u_0, v_0, w_0) &= -p \cos (n^\circ, X),\\f_y(u_0, v_0, w_0) &= -p \cos (n^\circ, Y),\\f_z(u_0, v_0, w_0) &= -p \cos (n^\circ, Z),\end{aligned} \tag{V.24}$$

$$\begin{aligned}f_x(u_0 + \alpha u_1, v_0 + \alpha v_1, w_0 + \alpha w_1) &= -p \cos (n^1, X),\\f_y(u_0 + \alpha u_1, v_0 + \alpha v_1, w_0 + \alpha w_1) &= -p \cos (n^1, Y),\\f_z(u_0 + \alpha u_1, v_0 + \alpha v_1, w_0 + \alpha w_1) &= -p \cos (n^1, Z),\end{aligned} \tag{V.25}$$

where p is the external pressure and n° and n^1 denote the normal directions to the bounding surfaces of the body in the first and second positions of equilibrium.

On the basis of the formulas of § 2, we have

$$
\begin{aligned}
\cos(n^\circ, X) &= \frac{1}{1+\overset{\circ}{E}_n}\left[\left(1+\frac{\partial u_0}{\partial x}\right)\cos(n, X)\right.\\
&\quad\left.+\frac{\partial u_0}{\partial y}\cos(n, Y)+\frac{\partial u_0}{\partial z}\cos(n, Z)\right],\\
\cos(n^\circ, Y) &= \frac{1}{1+\overset{\circ}{E}_n}\left[\frac{\partial v_0}{\partial x}\cos(n, X)\right.\\
&\quad\left.+\left(1+\frac{\partial v_0}{\partial y}\right)\cos(n, Y)+\frac{\partial v_0}{\partial z}\cos(n, Z)\right],\\
\cos(n^\circ, Z) &= \frac{1}{1+\overset{\circ}{E}_n}\left[\frac{\partial w_0}{\partial x}\cos(n, X)\right.\\
&\quad\left.+\frac{\partial w_0}{\partial y}\cos(n, Y)+\left(1+\frac{\partial w_0}{\partial z}\right)\cos(n, Z)\right],
\end{aligned}
\tag{V.26}
$$

$$
\begin{aligned}
\cos(n^1, X) &= \frac{1}{1+E_n^1}\left[\left(1+\frac{\partial u_0}{\partial x}+\alpha\frac{\partial u_1}{\partial x}\right)\cos(n, X)\right.\\
&\quad\left.+\left(\frac{\partial u_0}{\partial y}+\alpha\frac{\partial u_1}{\partial y}\right)\cos(n, Y)+\left(\frac{\partial u_0}{\partial z}+\alpha\frac{\partial u_1}{\partial z}\right)\cos(n, Z)\right],\\
\cos(n^1, Y) &= \frac{1}{1+E_n^1}\left[\left(\frac{\partial v_0}{\partial x}+\alpha\frac{\partial v_1}{\partial x}\right)\cos(n, X)\right.\\
&\quad\left.+\left(1+\frac{\partial v_0}{\partial y}+\alpha\frac{\partial v_1}{\partial y}\right)\cos(n, Y)+\left(\frac{\partial v_0}{\partial z}+\alpha\frac{\partial v_1}{\partial z}\right)\cos(n, Z)\right],\\
\cos(n^1, Z) &= \frac{1}{1+E_n^1}\left[\left(\frac{\partial w_0}{\partial x}+\alpha\frac{\partial w_1}{\partial x}\right)\cos(n, X)\right.\\
&\quad\left.+\left(\frac{\partial w_0}{\partial y}+\alpha\frac{\partial w_1}{\partial y}\right)\cos(n, Y)+\left(1+\frac{\partial w_0}{\partial z}+\alpha\frac{\partial w_1}{\partial z}\right)\cos(n, Z)\right],
\end{aligned}
\tag{V.27}
$$

where n denotes the normal direction to the bounding surface before the deformation, and $\overset{\circ}{E}_n$, E_n^1 are the relative elongations in that direction corresponding to the first and second positions of equilibrium.

The assumption that the strains are small enables us to simplify these formulas by neglecting $\overset{\circ}{E}_n$ and E_n^1 in comparison with unity.

Now, substituting (V.26) into (V.24) and (V.27) into (V.25), we obtain the following boundary conditions of the problem of elastic stability for a body subjected to hydrostatic pressure:

$$(\sigma'_{xx} - \omega'_z \overset{\circ}{\sigma}_{xy} + \omega'_y \overset{\circ}{\sigma}_{xz}) \cos(n, X) + (\sigma'_{xy} - \omega'_z \overset{\circ}{\sigma}_{yy} + \omega'_y \overset{\circ}{\sigma}_{yz}) \cos(n, Y) + (\sigma'_{xz} - \omega'_z \overset{\circ}{\sigma}_{yz} + \omega'_y \overset{\circ}{\sigma}_{zz}) \cos(n, Z)$$
$$= -p\left[\frac{\partial u_1}{\partial x}\cos(n, X) + \frac{\partial u_1}{\partial y}\cos(n, Y) + \frac{\partial u_1}{\partial z}\cos(n, Z)\right]. \qquad \text{(V.28)}$$

Furthermore, taking into account the formulas

$$\frac{\partial u_1}{\partial x} = e'_{xx} \qquad \frac{\partial u_1}{\partial y} = \frac{1}{2}e'_{xy} - \omega'_z,$$
$$\frac{\partial u_1}{\partial z} = \frac{1}{2}e'_{xz} + \omega'_y, \qquad \text{(V.29)}$$

and observing that in the preceding section it was found possible to neglect the parameters e'_{ij} in comparison with ω'_j, we arrive at the following final formulation of the boundary conditions (V.28):

$$(\sigma'_{xx} - \omega'_z \overset{0}{\sigma}_{xy} + \omega'_y \overset{0}{\sigma}_{xz}) \cos(n, X)$$
$$+ (\sigma'_{xy} - \omega'_z \overset{0}{\sigma}_{yy} + \omega_y \overset{0}{\sigma}_{yz}) \cos(n, Y)$$
$$+ (\sigma'_{xz} - \omega'_z \overset{0}{\sigma}_{yz} + \omega'_y \overset{0}{\sigma}_{zz}) \cos(n, Z) \qquad \text{(V.30)}$$
$$= p\,[\omega'_z \cos(n, Y) - \omega'_y \cos(n, Z)],$$

. . . .

§ 45. Energy Criterion for the Determination of Critical Loads

The above method reduces the problem of finding critical loads to that of determining the characteristic values of a system of linear homogeneous differential equations (V.16) with boundary conditions (V.19) and (V.23). This method has the advantage of yielding very accurate solutions. However, in many practical problems, the solutions cannot be obtained, and it is then more convenient to solve the problem of elasticstability by using a different criterion. This will now be discussed.

By (III.45) and Hooke's law, the strain energy of an elastic body can be written in the form

$$A=\frac{1}{2}\int\int\int(\sigma_{xx}\varepsilon_{xx}+\sigma_{yy}\varepsilon_{yy}+\sigma_{zz}\varepsilon_{zz}+\sigma_{xy}\varepsilon_{xy}+\sigma_{xz}\varepsilon_{xz}+\sigma_{yz}\varepsilon_{yz})\,dxdydz\,. \tag{V.31}$$

Since this formula will be used in the sequel only for small deformations, no distinction is made here between the generalized stresses $\overset{*}{\sigma}_{ij}$ and the true stresses σ_{ij}. Let us apply (V.31) to the second of the two adjacent positions of equilibrium of the body, i. e., to the position to which the displacements (V.1) correspond. Contrary to our procedure in the preceding sections, we shall retain not only terms of the first order (i. e., those multiplied by α) but also those of the second order (those multiplied by α^2).

With this degree of accuracy (the need for which will appear later), we obtain

$$A=A^{\circ}+\alpha A^{(1)}+\alpha^2A^{(2)}, \tag{V.32}$$

where

$$A^{\circ}=\frac{1}{2}\int\int\int(\overset{0}{\sigma}_{xx}\overset{0}{\varepsilon}_{xx}+\overset{0}{\sigma}_{yy}\overset{0}{\varepsilon}_{yy}+\overset{0}{\sigma}_{zz}\overset{0}{\varepsilon}_{zz}+\overset{0}{\sigma}_{xy}\overset{0}{\varepsilon}_{xy}+\overset{0}{\sigma}_{xz}\overset{0}{\varepsilon}_{xz}+\overset{0}{\sigma}_{yz}\overset{0}{\varepsilon}_{yz})\,dxdydz,$$

$$A^{(1)}=\frac{1}{2}\int\int\int(\overset{0}{\sigma}_{xx}\varepsilon'_{xx}+\overset{0}{\sigma}_{yy}\varepsilon'_{yy}+\overset{0}{\sigma}_{zz}\varepsilon'_{zz}+\overset{0}{\sigma}_{xy}\varepsilon'_{xy}+\overset{0}{\sigma}_{xz}\varepsilon'_{xz}+\overset{0}{\sigma}_{yz}\varepsilon'_{yz}+\sigma'_{xx}\overset{0}{\varepsilon}_{xx}+\sigma'_{yy}\overset{0}{\varepsilon}_{yy}+\sigma'_{zz}\overset{0}{\varepsilon}_{zz}+\sigma'_{xy}\overset{0}{\varepsilon}_{xy}+\sigma'_{xz}\overset{0}{\varepsilon}_{xz}+\sigma'_{yz}\overset{0}{\varepsilon}_{yz})\,dxdydz,$$

$$A^{(2)} = \frac{1}{2}\int\int\int(\sigma'_{xx}\varepsilon'_{xx} + \sigma'_{yy}\varepsilon'_{yy} + \sigma'_{zz}\varepsilon'_{yy} + \sigma'_{xy}\varepsilon'_{xy}$$

$$+ \sigma'_{xz}\varepsilon'_{xz} + \sigma'_{yz}\varepsilon'_{yz} + \overset{0}{\sigma}_{xx}\varepsilon''_{xx} + \overset{0}{\sigma}_{yy}\varepsilon''_{yy}$$

$$+ \overset{0}{\sigma}_{zz}\varepsilon''_{yy} + \overset{0}{\sigma}_{xy}\varepsilon''_{xy} + \overset{0}{\sigma}_{xz}\varepsilon''_{xz} + \overset{0}{\sigma}_{yz}\varepsilon''_{yz} + \sigma''_{xx}\overset{0}{\varepsilon}_{xx} + \sigma''_{yy}\overset{0}{\varepsilon}_{yy}$$

$$+ \sigma''_{zz}\overset{0}{\varepsilon}_{zz} + \sigma''_{xy}\overset{0}{\varepsilon}_{xy} + \sigma''_{xz}\overset{0}{\varepsilon}_{xz} + \sigma''_{yz}\overset{0}{\varepsilon}_{yz})\, dxdydz. \quad \text{(V.33)}$$

Here the parameters $\overset{0}{\varepsilon}_{ij}$, ε'_{ij}, ε''_{ij} are determined by (V.3), (V.4), (V.5) and the parameters $\overset{0}{\sigma}_{ij}$, σ'_{ij}, σ''_{ij} are related to them by formulas of the form (V.7).

In virtue of formulas (V.7), the expressions for $A^{(1)}$ and $A^{(2)}$ can be written more simply as follows:

$$A^{(1)} = \int\int\int(\overset{0}{\sigma}_{xx}\varepsilon'_{xx} + \overset{0}{\sigma}_{yy}\varepsilon'_{yy} + \overset{0}{\sigma}_{zz}\varepsilon'_{zz} + \overset{0}{\sigma}_{xy}\varepsilon'_{xy}$$

$$+ \overset{0}{\sigma}_{xz}\varepsilon'_{xz} + \overset{0}{\sigma}_{yz}\varepsilon'_{yz})\, dxdydz,$$

$$A^{(2)} = \frac{1}{2}\int\int\int[\sigma'_{xx}\varepsilon'_{xx} + \sigma'_{yy}\varepsilon'_{yy} + \sigma'_{zz}\varepsilon'_{zz} + \sigma'_{xy}\varepsilon'_{xy}$$

$$+ \sigma'_{xz}\varepsilon'_{xz} + \sigma'_{yz}\varepsilon'_{yz} + 2\,(\overset{0}{\sigma}_{xx}\varepsilon''_{xx} + \overset{0}{\sigma}_{yy}\varepsilon''_{yy}$$

$$+ \overset{0}{\sigma}_{zz}\varepsilon''_{zz} + \overset{0}{\sigma}_{xy}\varepsilon''_{xy} + \overset{0}{\sigma}_{xz}\varepsilon''_{xz} + \overset{0}{\sigma}_{yz}\varepsilon''_{yz})]\, dxdydz. \quad \text{(V.34)}$$

This is easily verified by expressing σ'_{ij} in terms of ε_{ij}, and $\overset{0}{\varepsilon}_{ij}$ in terms of $\overset{0}{\sigma}_{ij}$, in the last two equations of (V.33).

We now make use of the principle of virtual displacements. However, we shall vary only the increments αu_1, αv_1, αw_1, rather than the total displacements (V.1), i. e., we shall consider virtual displacements of the form $\alpha\delta u_1$, $\alpha\delta v_1$, $\alpha\delta w_1$. Here δu_1, δv_1, δw_1 must be taken equal to zero at those points of the bounding surface at which the displacements are prescribed.

In this formulation, the principle of virtual displacements for the second of the two positions of equilibrium under consideration is given by

$$\delta A = \alpha\,[\delta\,(A^{(1)}) + \alpha\delta\,(A^{(2)})] = \alpha\,[\delta\,(R_1) + \delta\,(R_2)], \quad \text{(V.35)}$$

where $\alpha\delta R_1$ is the work done by the body forces, and $\alpha\delta R_2$ is the work done by the surface forces through the virtual displacements $\alpha\delta u_1$, $\alpha\delta v_1$, $\alpha\delta w_1$.

Since A^0 is independent of u_1, v_1, w_1, the variation $\delta(A^0)=0$.

The work done by the body forces is equal to

$$\alpha\delta R_1=\alpha\int\int\int[F_\xi\delta u_1+F_\eta\delta v_1+F_\zeta\delta w_1]\,dxdydz, \quad \text{(V.36)}$$

where F_ξ, F_η, F_ζ denote the values of the projections of the specific body force at the second position of equilibrium. According to § 43, however, the magnitudes and directions of the body forces can be considered independent of the strain. This enables us to replace F_ξ, F_η, F_ζ in (V.36) by their values at the initial position of equilibrium.

The work done by the surface forces is determined by

$$\alpha\delta R_2=\alpha\int\int[f_\xi\delta u_1+f_\eta\delta v_1+f_\zeta\delta w_1]\,d\Omega, \quad \text{(V.37)}$$

where f_ξ, f_η, f_ζ are the components of the specific surface load at the second position of equilibrium. These cannot be replaced by the analogous components at the initial position of equilibrium (as has already been noted in the preceding section).

As a result, (V.37) may be written in the form

$$\delta R_2=\delta R_2^{(1)}+\alpha\delta R_2^{(2)}, \quad \text{(V.38)}$$

where

$$\delta R_2^{(1)}=\int\int[f_\xi(u_0, v_0, w_0)\,\delta u_1+f_\eta(u_0, v_0, w_0)\,\delta v_1 +f_\zeta(u_0, v_0, w_0)\,\delta w_1]\,d\Omega=\int\int[f_\xi^0\delta u_1+f_\eta^0\delta v_1 +f_\zeta^0\delta w_1]\,d\Omega, \quad \text{(V.39)}$$

$$\delta R_2^{(2)}=\lim_{\alpha\to 0}\left\{\frac{1}{\alpha}\int\int[(f_\xi-f_\xi^0)\,\delta u_1+(f_\eta-f_\eta^0)\,\delta v_1 +(f_\zeta-f_\zeta^0)\,\delta w_1]\,d\Omega\right\}.$$

It follows from the above that (V.35), after cancellation of the infinitesimal α, becomes

$$\delta\,[A^{(1)}]+\alpha\delta\,[A^{(2)}]=\delta R_1+\delta R_2^{(1)}+\alpha\delta R_2^{(2)}. \quad \text{(V.40)}$$

Observe now that

$\delta A^{(1)}$,because of the first equation of (V.34), is the variation in the strain energy of the body at the initial equilibrium position resulting from the virtual displacements δu_1, δv_1, δw_1;

δR_1 is the work done by the body forces which correspond to the initial position of equilibrium (through the same virtual displacements);

$\delta R_2^{(1)}$ is the work done by the surface forces at the initial position of equilibrium (through the same virtual displacements).

Hence, since the position of the body which is characterized by the displacements u_0, v_0, w_0 is a position of equilibrium, we have

$$\delta\,[A^{(1)}]=\delta R_1+\delta k_2^{(1)}, \quad \text{(V.41)}$$

which is merely the mathematical formulation of the principle of virtual displacements at the first of the two positions of equilibrium examined.

After canceling α from (V.40), it follows that

$$\delta\,[A^{(2)}]=\delta R_2^{(2)}. \quad \text{(V.42)}$$

This last equation is the variational formulation of the problem of elastic stability. The differential equations of the form (V.10) and the corresponding boundary conditions can be derived from it by making use of operations analogous to those of the third chapter. Similarly, the system (V.9) can be obtained from (V.41). We leave the calculations to the reader.

In the above derivation all terms were retained, except those which could be neglected because of the smallness of elongations and shears. However, in the two preceding sections it was established that, in most practical situations, the initial state of stress of the body could be considered as a problem of the classical theory. This enabled us to replace the equations (V.9) by the system (V.11), and the equations (V.10) by the system (V.16), which is equivalent to adopting the equations (V.15) and neglecting the parameters e'_{ij} as compared with ω'_j.

In that case, in the second of formulas (V.34), ε'_{ij} can be replaced by e'_{ij} with the corresponding change in the formulas of the form (V.7) which express σ'_{ij} in terms of ε'_{ij}. Furthermore, the parameters ε''_{ij} of (V.5) can be represented in the form

$$\varepsilon''_{xx} = \frac{1}{2}\left[(e'_{xx})^2 + \left(\frac{1}{2}e'_{xy} + \omega'_z\right)^2 + \left(\frac{1}{2}e'_{xz} - \omega'_y\right)^2\right],$$

$$\cdots\cdots\cdots\cdots\cdots\cdots\cdots\cdots$$

$$\varepsilon''_{xy} = e'_{xx}\left(\frac{1}{2}e'_{xy} - \omega'_z\right) + e'_{yy}\left(\frac{1}{2}e'_{xy} + \omega'_z\right) + \left(\frac{1}{2}e'_{xz} - \omega'_y\right)\left(\frac{1}{2}e'_{yz} + \omega'_x\right), \tag{V.43}$$

$$\cdots\cdots\cdots\cdots\cdots\cdots\cdots\cdots$$

Taking into account the fact that e'_{ij} can be neglected in comparison with ω'_j, as indicated above, (V.43) may be written more simply as follows:

$$\begin{aligned} \varepsilon''_{xx} &= \frac{1}{2}[(\omega'_y)^2 + (\omega'_z)^2], & \varepsilon''_{xy} &= -\omega'_x\omega'_y, \\ \varepsilon''_{yy} &= \frac{1}{2}[(\omega'_z)^2 + (\omega'_x)^2], & \varepsilon'_{xz} &= -\omega'_x\omega'_z, \\ \varepsilon''_{zz} &= \frac{1}{2}[(\omega'_x)^2 + (\omega'_y)^2], & \varepsilon''_{yz} &= -\omega'_y\omega'_z. \end{aligned} \tag{V.44}$$

Hence, $A^{(2)}$ can be expressed by the approximate formula

$$A^{(2)} = \frac{E}{2(1+\mu)} \int\int\int \left\{ \frac{1-\mu}{1-2\mu} (b_2')^2 - 2b_1' \right\} dx\ dy\ dz$$
$$+ \frac{1}{2} \int\int\int \left\{ \overset{\circ}{\sigma}_{xx} (\omega_y'^2 + \omega_z'^2) + \overset{\circ}{\sigma}_{yy} (\omega_x'^2 + \omega_z'^2) \right. \qquad \text{(V.45)}$$
$$+ \overset{\circ}{\sigma}_{zz} (\omega_x'^2 + \omega_y'^2) - 2 [\overset{\circ}{\sigma}_{xy} \omega_x' \omega_y' + \overset{\circ}{\sigma}_{xz} \omega_x' \omega_z'$$
$$\left. + \overset{\circ}{\sigma}_{yz} \omega_y' \omega_z'] \right\} dx\ dy\ dz,$$

where

$$b_2' = e_{xx}' + e_{yy}' + e_{zz}',$$
$$b_1' = e_{xx}' e_{yy}' + e_{xx}' e_{zz}' + e_{yy}' e_{zz}' - \frac{1}{2} (e_{xy}'^2 + e_{xz}'^2 + e_{yz}'^2).$$

This should be used in all those cases in which the initial state of stress is subject to the classical theory.

The formulation of the problem of elastic stability as given by (V.42) enables one to use the direct methods of the calculus of variations in the solution of concrete problems. This consists in representing u_1, v_1, w_1 in the form

$$u_1 = \alpha_1 u^{\mathrm{I}} + \alpha_2 u^{\mathrm{II}} + \alpha_3 u^{\mathrm{III}} + \ldots,$$
$$v_1 = \beta_1 v^{\mathrm{I}} + \beta_2 v^{\mathrm{II}} + \beta_3 v^{\mathrm{III}} + \ldots, \qquad \text{(V.46)}$$
$$w_1 = \gamma_1 w^{\mathrm{I}} + \gamma_2 w^{\mathrm{II}} + \gamma_3 w^{\mathrm{III}} + \ldots,$$

where α_j, β_j, γ_j are unknown coefficients independent of x, y, z and u^j, v^j, w^j are known functions of x, y, z which vanish at those boundary points at which the displacements are prescribed explicitly. The choice of these functions is to a great extent arbitrary. It is only important that the first terms in the series (V.46), i. e., the functions u^{I}, v^{I}, w^{I}, shall roughly correspond in character to the expected solutions (which, with some experience, it is always possible to predict qualitatively). Then the remaining terms of the series (V.46) are in the nature of corrections.

Since the functions u^j, v^j, w^j are known, the variation of the expressions (V.46) reduces to the variation of the coefficients α_j, β_j, γ_j which enter into them. Substituting (V.46) into (V.42) and requiring that the resulting equation be satisfied for arbitrary values of $\delta(\alpha_j)$, $\delta(\beta_j)$, $\delta(\gamma_j)$, we obtain a system of $3n$ linear homogeneous algebraic equations, where n is the number of terms retained in each of the series (V.46). This system has a nontrivial solution only if the determinant of the coefficients vanishes. These coefficients are functions, first of the parameters which characterize the dimensions of the body; second, of the parameters which characterize its elastic properties; and third, of the parameter of the external load.

The last of these parameters is the only unknown quantity in the characteristic determinant. The values of this parameter for which the determinant vanishes correspond, as is clear from the above, to the critical values of the load.

This approximate method for solving problems of the theory of elastic stability is among the most effective, since in many cases the retention of merely the first terms in the series (V.46) already yields an accurate result. At the same time, this method requires a certain experience on the part of those who use it, and requires primarily an understanding of the physical nature of the problem to be solved, since the accuracy of the method depends to a large extent on a successful intuitive choice of the basic approximating functions u^{I}, v^{I}, w^{I}.

CHAPTER VI

ON THE DEFORMATION OF FLEXIBLE BODIES

§ 46. Deformation of Plates

We shall discuss in the present chapter the specific simplifications which are possible in the study of the deformation of flexible bodies. We begin with an investigation of the deformation of plates, since this problem is the simplest.

Let us consider a thin plate of constant thickness δ bounded by a contour of arbitrary shape.

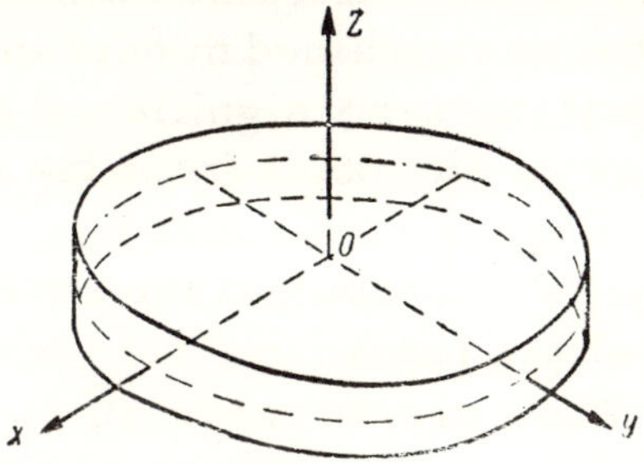

Fig. 17

We shall put the origin of a Cartesian coordinate system at some point of the middle plane of the plate, directing the Z-axis perpendicular to the plane and taking any two perpendicular straight lines in the middle plane as the X- and Y-axes (Fig. 17).

In developing the theory of deformation of thin plates which are subjected to a transverse load it is ordinarily supposed that every point of the plate remains after the deformation on the same perpendicular to the middle surface as before and that the distance of every point of the plate from the middle surface remains unchanged by the deformation.

These assumptions can be formulated analytically by means of the following three differential equations:

$$\begin{aligned}
\varepsilon_{xz} &= \left(1+\frac{\partial u}{\partial x}\right)\frac{\partial u}{\partial z}+\frac{\partial v}{\partial x}\frac{\partial v}{\partial z}+\frac{\partial w}{\partial x}\left(1+\frac{\partial w}{\partial z}\right)=0,\\
\varepsilon_{yz} &= \frac{\partial u}{\partial y}\frac{\partial u}{\partial z}+\left(1+\frac{\partial v}{\partial y}\right)\frac{\partial v}{\partial z}+\frac{\partial w}{\partial y}\left(1+\frac{\partial w}{\partial z}\right)=0, \qquad \text{(VI.1)}\\
\varepsilon_{zz} &= \frac{\partial w}{\partial z}+\frac{1}{2}\left[\left(\frac{\partial u}{\partial z}\right)^2+\left(\frac{\partial v}{\partial z}\right)^2+\left(\frac{\partial w}{\partial z}\right)^2\right]=0\,.
\end{aligned}$$

Setting ε_{xz} and ε_{yz} equal to zero in these equations means that those fibers of the plate which were perpendicular to the unstrained middle plane remain normal to the strained middle surface, while setting ε_{zz} equal to zero signifies the invariance of the distance of any point from the middle surface.

Let us try as solutions of the equations (VI.1) expressions of the form

$$\begin{aligned}
u &= \hat{u}(x, y)+z\vartheta(x, y),\\
v &= \hat{v}(x, y)+z\psi(x, y), \qquad \text{(VI.2)}\\
w &= \hat{w}(x, y)+z\chi(x, y),
\end{aligned}$$

where $\hat{u}$, $\hat{v}$, $\hat{w}$ denote the displacements of the points of the middle plane (which follows by setting $z=0$ in (VI.2)).

Substituting (VI.2) into (VI.1), we obtain

$$\begin{aligned}
&\left(1+\frac{\partial \hat{u}}{\partial x}\right)\vartheta+\frac{\partial \hat{v}}{\partial x}\psi+\frac{\partial \hat{w}}{\partial x}(1+\chi)=0,\\
&\frac{\partial \hat{u}}{dy}\vartheta+\left(1+\frac{\partial \hat{v}}{\partial y}\right)\psi+\frac{\partial \hat{w}}{\partial y}(1+\chi)=0, \qquad \text{(VI.3)}\\
&\vartheta^2+\psi^2+(1+\chi)^2=1,
\end{aligned}$$

which enable us to express ϑ, ψ , and χ in terms of the displacements of the middle plane, $\hat{u}$, $\hat{v}$, $\hat{w}$.

If we regard the first two of the equations (VI.3) as an algebraic system in the unknowns ϑ and ψ, we find

$$\vartheta = \frac{\hat{\alpha}_{31}}{\hat{\alpha}_{33}}(1+\chi), \quad \psi = \frac{\hat{\alpha}_{32}}{\hat{\alpha}_{33}}(1+\chi), \tag{VI.4}$$

where $\hat{\alpha}_{31}$, $\hat{\alpha}_{32}$, $\hat{\alpha}_{33}$ are obtained from equations (I.15) by substituting $\hat{u}$, $\hat{v}$, $\hat{w}$ for u, v, w.

Introducing (VI.4) into the third of equations (VI.3), we get

$$\chi = \frac{\hat{\alpha}_{33}}{\sqrt{\hat{\alpha}_{31}^2 + \hat{\alpha}_{32}^2 + \hat{\alpha}_{33}^2}} - 1, \tag{VI.5}$$

but, in virtue of Table II of § 2,

$$\sqrt{\alpha_{31}^2 + \alpha_{32}^2 + \alpha_{33}^2} = \frac{D}{1+E_\zeta}, \tag{VI.6}$$

where $D = 1 + \Delta$ and E_ζ is the relative elongation of a line element of the plate which is parallel to the Z-axis after the deformation. Substitution in (VI.6) of $\hat{\alpha}_{ij}$ for α_{ij}, which corresponds to the replacement of u, v, w by $\hat{u}$, $\hat{v}$, $\hat{w}$, is equivalent to considering D and E_ζ on the middle plane.

Hence it is seen that if the shears and elongations are neglected in comparison with unity, one obtains

$$\sqrt{\hat{\alpha}_{31}^2 + \hat{\alpha}_{32}^2 + \hat{\alpha}_{33}^2} \approx 1 \tag{VI.7}$$

and consequently (to the same degree of accuracy),

$$\begin{aligned} \vartheta &\approx \hat{\alpha}_{31} = -\frac{\partial \hat{w}}{\partial x}\left(1 + \frac{\partial \hat{v}}{\partial y}\right) + \frac{\partial \hat{v}}{\partial x}\frac{\partial \hat{w}}{\partial y}, \\ \psi &\approx \hat{\alpha}_{32} = -\frac{\partial \hat{w}}{\partial y}\left(1 + \frac{\partial \hat{u}}{\partial x}\right) + \frac{\partial \hat{u}}{\partial y}\frac{\partial \hat{w}}{\partial x}, \\ \chi &\approx \hat{\alpha}_{33} - 1 = \frac{\partial \hat{u}}{\partial x} + \frac{\partial \hat{v}}{\partial y} + \frac{\partial \hat{u}}{\partial x}\frac{\partial \hat{v}}{\partial y} - \frac{\partial \hat{u}}{\partial y}\frac{\partial \hat{v}}{\partial x}. \end{aligned} \tag{VI.8}$$

If these values of ϑ, ψ, and χ are substituted in (VI.2), the displacements of an arbitrary point of the plate are obtained in terms of the displacements $\hat{u}$, $\hat{v}$, $\hat{w}$ (functions of x and y alone) of the corresponding point of the middle plane.

The law of variation of the displacements along the thickness of the plate turns out to be linear.

Let us observe that when we assumed the elongations and shears to be negligible compared to unity, we imposed no restrictions on the magnitude of the rotations. Hence the displacements which are given by (VI.2) and (VI.8) can be called displacements corresponding to strong bending of the plate.

Substituting these values of the displacements into the general expressions for the strain components ε_{xx}, ε_{yy}, ε_{xy}, we find

$$\begin{aligned} \varepsilon_{xx} &= \hat{\varepsilon}_{xx} + z\,\varkappa_{xx} + z^2\,\nu_{xx}, \\ \varepsilon_{yy} &= \hat{\varepsilon}_{yy} + z\,\varkappa_{yy} + z^2\,\nu_{yy}, \\ \varepsilon_{xy} &= \hat{\varepsilon}_{xy} + z\,\varkappa_{xy} + z^2\,\nu_{xy}, \end{aligned} \tag{VI.9}$$

where

$$\begin{aligned} \hat{\varepsilon}_{xx} &= \frac{\partial \hat{u}}{\partial x} + \frac{1}{2}\left[\left(\frac{\partial \hat{u}}{\partial x}\right)^2 + \left(\frac{\partial \hat{v}}{\partial x}\right)^2 + \left(\frac{\partial \hat{w}}{\partial x}\right)^2\right], \\ \hat{\varepsilon}_{yy} &= \frac{\partial \hat{v}}{\partial y} + \frac{1}{2}\left[\left(\frac{\partial \hat{u}}{\partial y}\right)^2 + \left(\frac{\partial \hat{v}}{\partial y}\right)^2 + \left(\frac{\partial \hat{w}}{\partial y}\right)^2\right] \\ \hat{\varepsilon}_{xy} &= \frac{\partial \hat{u}}{\partial y} + \frac{\partial \hat{v}}{\partial x} + \frac{\partial \hat{u}}{\partial y}\frac{\partial \hat{u}}{\partial x} + \frac{\partial \hat{v}}{\partial x}\frac{\partial \hat{v}}{\partial y} + \frac{\partial \hat{w}}{\partial x}\frac{\partial \hat{w}}{\partial y} \end{aligned}$$

$$\begin{aligned} \varkappa_{xx} &= \frac{\partial \vartheta}{\partial x} + \frac{\partial \hat{u}}{\partial x}\frac{\partial \vartheta}{\partial x} + \frac{\partial \hat{v}}{\partial x}\frac{\partial \psi}{\partial x} + \frac{\partial \hat{w}}{\partial x}\frac{\partial \chi}{\partial x}, \\ \varkappa_{yy} &= \frac{\partial \psi}{\partial y} + \frac{\partial \hat{u}}{\partial y}\frac{\partial \vartheta}{\partial y} + \frac{\partial \hat{v}}{\partial y}\frac{\partial \psi}{\partial y} + \frac{\partial \hat{w}}{\partial y}\frac{\partial \chi}{\partial y}, \\ \varkappa_{xy} &= \frac{\partial \vartheta}{\partial y} + \frac{\partial \psi}{\partial x} + \frac{\partial \hat{u}}{\partial x}\frac{\partial \vartheta}{\partial y} + \frac{\partial \hat{u}}{\partial y}\frac{\partial \vartheta}{\partial x} \\ &\quad + \frac{\partial \hat{v}}{\partial x}\frac{\partial \psi}{\partial y} + \frac{\partial \hat{v}}{\partial y}\frac{\partial \psi}{\partial x} + \frac{\partial \hat{w}}{\partial x}\frac{\partial \chi}{\partial y} + \frac{\partial \hat{w}}{\partial y}\frac{\partial \chi}{\partial x}, \end{aligned} \tag{VI.10}$$

$$\nu_{xx}=\frac{1}{2}\left[\left(\frac{\partial\vartheta}{\partial x}\right)^2+\left(\frac{\partial\psi}{\partial x}\right)^2+\left(\frac{\partial\chi}{\partial x}\right)^2\right],$$
$$\nu_{yy}=\frac{1}{2}\left[\left(\frac{\partial\vartheta}{\partial y}\right)^2+\left(\frac{\partial\psi}{\partial y}\right)^2+\left(\frac{\partial\chi}{\partial y}\right)^2\right], \qquad \text{(VI.11)}$$
$$\nu_{xy}=\frac{\partial\vartheta}{\partial x}\frac{\partial\vartheta}{\partial y}+\frac{\partial\psi}{\partial x}\frac{\partial\psi}{\partial y}+\frac{\partial\chi}{\partial x}\frac{\partial\chi}{\partial y}.$$

For $z=0$ equations (VI.9) become

$$\varepsilon_{xx}=\hat{\varepsilon}_{xx},\ \varepsilon_{yy}=\hat{\varepsilon}_{yy},\varepsilon_{xy}=\hat{\varepsilon}_{xy},$$

whence it is apparent that $\hat{\varepsilon}_{xx}$, $\hat{\varepsilon}_{yy}$, $\hat{\varepsilon}_{xy}$ are the elongations and shear of the middle plane of the plate.

On the other hand, the terms $z\varkappa_{xx}$, $z\varkappa_{yy}$, $z\varkappa_{xy}$ in (VI.9) correspond to the elongations and shears which vary linearly along the thickness of the plate, i. e., correspond to the bending and torsion of the plate. Finally, there are terms in (VI.9) containing z^2. Their presence indicates that to the linear law of variation of displacements along the thickness of the plate there generally corresponds a nonlinear variation in the strain components ε_{xx}, ε_{yy}, ε_{xy}. However, for small elongations and shears the corrections introduced into (VI.9) by the terms $\nu_{xx}z^2$, $\nu_{yy}z^2$, $\nu_{xy}z^2$ are insignificant, as will be shown below by an example.

Thus, one can write

$$\varepsilon_{xx}=\hat{\varepsilon}_{xx}+z\varkappa_{xx},$$
$$\varepsilon_{yy}=\hat{\varepsilon}_{yy}+z\varkappa_{yy}, \qquad \text{(VI.12)}$$
$$\varepsilon_{xy}=\hat{\varepsilon}_{xy}+z\varkappa_{xy}.$$

The parameters $\varkappa_{xx}$, $\varkappa_{yy}$, $\varkappa_{xy}$ characterize the curvature of the strained middle surface of the plate.

When the angles of rotation of the elements of the plate are negligibly small compared to unity, the formulas derived admit of an essential simplification. Indeed, in that special case, the derivatives $\frac{\partial \hat{u}}{\partial x}, \frac{\partial \hat{v}}{\partial y}$, in accordance with § 14, can be regarded as magnitudes of the same order as the elongations $\hat{\varepsilon}_{xx}$, $\hat{\varepsilon}_{yy}$ of the middle surface. Furthermore, in this case, the derivatives $\frac{\partial \hat{u}}{\partial y}, \frac{\partial \hat{v}}{\partial x}$ must be regarded as magnitudes of the same order as the shear $\hat{\varepsilon}_{xy}$. This follows from the fact that, for small deformations the plate, being a massive body in its plane, does not admit of relative rotations of its elements about the Z-axis which are large compared to the shears.

In accordance with these considerations, one can omit all nonlinear terms which contain the derivatives of the displacements $\hat{u}$ and $\hat{v}$ in (VI.10) (bearing in mind the special case that the angles of rotation of the elements of the plate are small compared to unity).

With this simplification, we arrive at the expressions

$$\begin{aligned} \varepsilon_{xx} &= \frac{\partial \hat{u}}{\partial x} + \frac{1}{2}\left(\frac{\partial \hat{w}}{\partial x}\right)^2, \quad \varepsilon_{yy} = \frac{\partial \hat{v}}{\partial y} + \frac{1}{2}\left(\frac{\partial \hat{w}}{\partial y}\right)^2, \\ \hat{\varepsilon}_{xy} &= \frac{\partial \hat{u}}{\partial y} + \frac{\partial \hat{v}}{\partial x} + \frac{\partial \hat{w}}{\partial x}\frac{\partial \hat{w}}{\partial y}, \\ \varkappa_{xx} &= \frac{\partial \vartheta}{\partial x} \approx -\frac{\partial^2 \hat{w}}{\partial x^2}, \quad \varkappa_{yy} = \frac{\partial \psi}{\partial y} \approx -\frac{\partial^2 \hat{w}}{\partial y^2}, \\ \varkappa_{xy} &= \frac{\partial \vartheta}{\partial y} + \frac{\partial \psi}{\partial x} \approx -2\frac{\partial^2 \hat{w}}{\partial x \partial y}. \end{aligned} \tag{VI.13}$$

Substitution of (VI.13) into (VI.12) leads to the formulas for the deformation of plates proposed by T. v. Kármán. To this degree of accuracy the displacements of the points of the plate are given by

$$u = \hat{u} - z\frac{\partial \hat{w}}{\partial x}, \quad v = \hat{v} - z\frac{\partial \hat{w}}{\partial y}, \quad w = \hat{w}. \tag{VI.14}$$

The greatest conceivable simplification of the formulas (VI.13) consists in neglecting all the nonlinear terms in them. As a result, one obtains the formulas of the classical theory of plates. It is seen from the above that, generally speaking, it is necessary for the squares of the rotations

$$\omega_x \approx \frac{\partial \hat{w}}{\partial y}, \quad \omega_y \approx -\frac{\partial \hat{w}}{\partial x}$$

to be negligible compared to the elongations and shears in order that the classical theory be applicable. Thus, the classical theory can be used only in the analysis of plates under weak bending. The strong bending of plates, as has already been mentioned, is described by formulas (VI.10), (VI.8), and (VI.12). v. Kármán's equations (VI.13) refer to the intermediate case.

§ 47. Two-dimensional Deformation of an Infinitely Long Strip

Let us suppose that a plate has the form of an infinitely long strip with straight parallel edges. We shall assume that its deformation is independent of y and, further, that the displacement v is equal to zero.

The plane problem formulated in this manner is essentially equivalent to the problem of the bending of a beam. The equations of the last section, as applied to this problem, are

$$\varepsilon_{yy} = \varepsilon_{xy} = 0,$$
$$\varepsilon_{xx} = \hat{\varepsilon}_{xx} + z\varkappa_{xx} + z^2 \nu_{xx}, \tag{VI.15}$$

where

$$\hat{\varepsilon}_{xx} = \frac{d\hat{u}}{dx} + \frac{1}{2}\left[\left(\frac{d\hat{u}}{dx}\right)^2 + \left(\frac{d\hat{w}}{dx}\right)^2\right],$$
$$\varkappa_{xx} = \left(1 + \frac{d\hat{u}}{dx}\right)\frac{d\vartheta}{dx} + \frac{d\hat{w}}{dx}\frac{d\chi}{dx},$$

$$\nu_{xx}=\frac{1}{2}\left[\left(\frac{d\vartheta}{dx}\right)^2+\left(\frac{d\chi}{dx}\right)^2\right], \tag{VI.16}$$
$$\vartheta=-\frac{d\hat{w}}{dx},\quad \chi=\frac{d\hat{u}}{dx}.$$

Since the derivatives $\frac{d\hat{u}}{dx}$, $\frac{d\hat{w}}{dx}$ are considered large as compared to the elongation $\hat{\varepsilon}_{xx}$, the right-hand side of the first of equations (VI.16) is a small difference of large terms and, in determining the relation between $\frac{d\hat{u}}{dx}$ and $\frac{d\hat{w}}{dx}$, one can approximately set $\hat{\varepsilon}_{xx}=0$. Then

$$1+\frac{d\hat{u}}{dx}=\sqrt{1-\left(\frac{d\hat{w}}{dx}\right)^2}, \tag{VI.17}$$

whence

$$\frac{d\chi}{dx}=-\frac{1}{\sqrt{1-\left(\frac{d\hat{w}}{dx}\right)^2}}\frac{d\hat{w}}{dx}\frac{d^2\hat{w}}{dx^2}$$
$$\varkappa_{xx}=-\frac{1}{\sqrt{1-\left(\frac{d\hat{w}}{dx}\right)^2}}\frac{d^2\hat{w}}{dx^2}, \tag{VI.18}$$
$$\nu_{xx}=\frac{1}{1-\left(\frac{d\hat{w}}{dx}\right)^2}\frac{1}{2}\left(\frac{d^2\hat{w}}{dx^2}\right)^2.$$

Substituting these values of $\varkappa_{xx}$, ν_{xx} in the last of equations (VI.15), we find

$$\varepsilon_{xx}=\hat{\varepsilon}_{xx}+z\,\varkappa_{xx}\left(1+\frac{1}{2}z\,\varkappa_{xx}\right). \tag{VI.19}$$

Since $z\varkappa_{xx}$ is of the same order as the elongations, which are assumed to be negligible compared to unity, we have

$$\varepsilon_{xx}\approx\hat{\varepsilon}_{xx}+z\,\varkappa_{xx}=\hat{\varepsilon}_{xx}-\frac{z}{\sqrt{1-\left(\frac{d\hat{w}}{dx}\right)^2}}\frac{d^2\hat{w}}{dx^2}. \tag{VI.20}$$

We shall now interpret the parameter $\varkappa_{xx}$. To this end, consider Fig. 18.

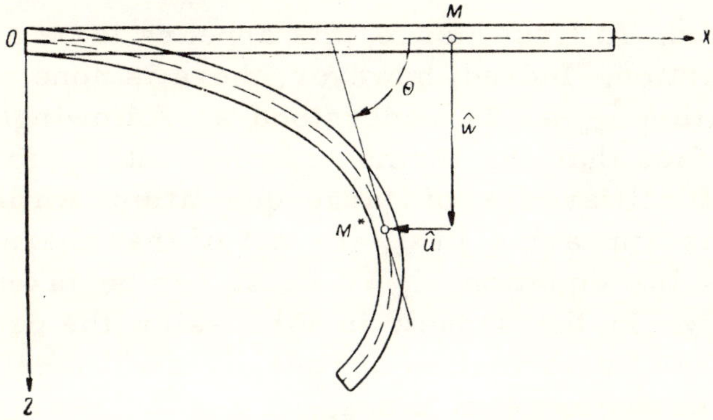

Fig. 18

The Cartesian coordinate x which determines the position of a point on the middle plane of the plate before the deformation becomes, after the deformation, the length of the arc of that cylindrical surface into which the middle plane is transformed. Hence it is clear that

$$\frac{d\hat{w}}{dx} = \sin\,\theta,$$

whence

$$\varkappa_{xx} = -\frac{1}{\sqrt{1-\left(\frac{d\hat{w}}{dx}\right)^2}}\frac{d^2\hat{w}}{dx^2} = -\frac{d\theta}{dx} = -\frac{d\theta}{ds}, \qquad \text{(VI.21)}$$

i. e., $\varkappa_{xx}$ is the curvature of the cylindrical surface into which the middle plane of the plate is deformed. This result, together with (VI.20), shows that our deductions coincide completely with the usual theory of the plane deformation of flexible, initially straight rods.

In conclusion, we must call attention to the fact that when we assumed the equation $\hat{\varepsilon}_{xx}=0$, in simplifying the equation for $\varkappa_{xx}$, we nevertheless retained $\hat{\varepsilon}_{xx}$ in equation (VI.20). This was done because the first term of (VI.20) can, under certain conditions, be no less significant than the second. At first glance, this appears to be a contradiction. Indeed, however, there is none, if the equation $\hat{\varepsilon}_{xx}=0$ is understood as following from the fact that the terms entering into $\hat{\varepsilon}_{xx}$ form a small difference of large quantities, while the terms appearing in $\varkappa_{xx}$ are not of that character. Thus the equation $\hat{\varepsilon}_{xx}=0$ must not be taken literally. In the sequel, in discussing the general theory of deformation of rods, we shall meet with this device several times.

§ 48. Deformation of Shells

The theory of deformation of thin shells can be developed by analogy with the theory of deformation of plates, since shells are plates whose middle surface in the undeformed state is not a plane, but a curved surface.

We determine the position of points on the middle surface of the shell by the Gaussian curvilinear coordinates α_1 and α_2 of the surface, i. e., coordinates for which the coordinate lines are lines of principal curvature of the middle surface. The Lamé coefficients corresponding to this curvilinear system are denoted by A_1 and A_2 and the principal radii of curvature by R_1 and R_2.

Starting with this two-dimensional system of curvilinear coordinates, we shall construct a three-dimensional system. To this end, we drop a perpendicular to the middle surface from an arbitrary point M not lying on the surface. Then the position of the point M in space is fixed by three parameters: the coordinates α_1 and α_2 of the base of the perpendicular and the length z of the perpendicular. We shall agree to consider z positive if the point lies on the side of the centers of negative curvature of the middle surface (if the points of the surface are of elliptic type, then z is considered positive if the point M lies on the convex side of the surface). In the contrary case, z is taken to be negative. As is clear from its construction, this system of coordinates α_1, α_2, z forms a triply orthogonal system. Its Lamé coefficients are

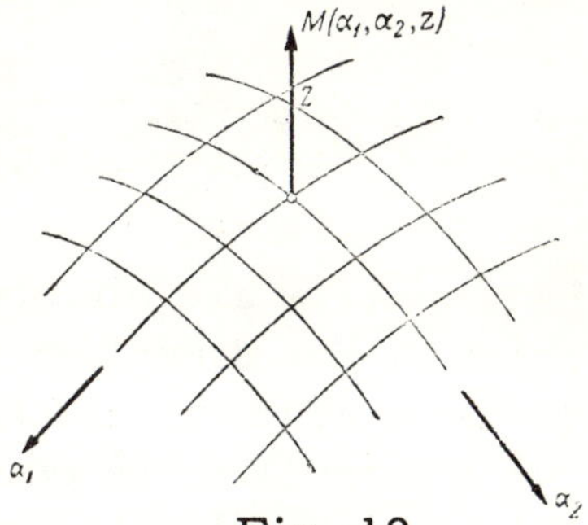

Fig. 19

$$H_1 = A_1\left(1+\frac{z}{R_1}\right), \quad H_2 = A_2\left(1+\frac{z}{R_2}\right), \quad H_3 = 1, \qquad \text{(VI.22)}$$

where A_1, A_2, R_1, R_2 (being functions of α_1 and α_2 alone) must satisfy the well-known Gauss-Codazzi relations of surface theory:

$$\frac{\partial}{\partial\alpha_1}\left(\frac{A_2}{R_2}\right) = \frac{1}{R_1}\frac{\partial A_2}{\partial\alpha_1}, \quad \frac{\partial}{\partial\alpha_2}\left(\frac{A_1}{R_1}\right) = \frac{1}{R_2}\frac{\partial A_1}{\partial\alpha_2},$$

$$\frac{\partial}{\partial\alpha_1}\left(\frac{1}{A_1}\frac{\partial A_2}{\partial\alpha_1}\right) + \frac{\partial}{\partial\alpha_2}\left(\frac{1}{A_2}\frac{\partial A_1}{\partial\alpha_2}\right) = -\frac{A_1A_2}{R_1R_2}. \qquad \text{(VI.23)}$$

Substituting the above expressions for the Lamé coefficients into equations (I.122) we obtain, with the aid of (VI.23), the relations

$$e_{11} = \frac{1}{1+\frac{z}{R_1}}\left(\frac{1}{A_1}\frac{\partial u}{\partial\alpha_1} + \frac{1}{A_1A_2}\frac{\partial A_1}{\partial\alpha_2}v + \frac{w}{R_1}\right),$$

$$e_{22} = \frac{1}{1+\frac{z}{R_2}}\left(\frac{1}{A_2}\frac{\partial v}{\partial\alpha_2} + \frac{1}{A_1A_2}\frac{\partial A_2}{\partial\alpha_1}u + \frac{w}{R_2}\right),$$

$$e_{zz} = \frac{\partial w}{\partial z}, \qquad \text{(VI.24)}$$

$$e_{12} = \frac{1}{1+\frac{z}{R_1}}\left(\frac{1}{A_1}\frac{\partial v}{\partial\alpha_1} - \frac{1}{A_1A_2}\frac{\partial A_1}{\partial\alpha_2}u\right) + \frac{1}{1+\frac{z}{R_2}}\left(\frac{1}{A_2}\frac{\partial u}{\partial\alpha_2} - \frac{1}{A_1A_2}\frac{\partial A_2}{\partial\alpha_1}v\right)$$

$$e_{1z}=\frac{\partial u}{\partial z}+\frac{1}{1+\frac{z}{R_1}}\left(\frac{1}{A_1}\frac{\partial w}{\partial \alpha_1}-\frac{u}{R_1}\right),$$

$$e_{2z}=\frac{\partial v}{\partial z}+\frac{1}{1+\frac{z}{R_2}}\left(\frac{1}{A_2}\frac{\partial w}{\partial \alpha_2}-\frac{v}{R_2}\right).$$

Here u, v, w are displacements of an arbitrary point in the direction of the lines α_1 and α_2 and of the normal to the middle surface.

Furthermore, introducing the Lamé coefficients (VI.22) into (I.124), we arrive at the following expressions for the components of the vector $\vec{\omega}$ along the indicated directions:

$$2\omega_1=-\frac{\partial v}{\partial z}+\frac{1}{1+\frac{z}{R_2}}\left(\frac{1}{A_2}\frac{\partial w}{\partial \alpha_2}-\frac{v}{R_2}\right),$$

$$2\omega_2=+\frac{\partial u}{\partial z}-\frac{1}{1+\frac{z}{R_1}}\left(\frac{1}{A_1}\frac{\partial w}{\partial \alpha_1}-\frac{u}{R_1}\right), \qquad \text{(VI.25)}$$

$$2\omega_z=\frac{1}{1+\frac{z}{R_1}}\left(\frac{1}{A_1}\frac{\partial v}{\partial \alpha_1}-\frac{1}{A_1A_2}\frac{\partial A_1}{\partial \alpha_2}u\right)-$$

$$-\frac{1}{1+\frac{z}{R_2}}\left(\frac{1}{A_2}\frac{\partial u}{\partial \alpha_2}-\frac{1}{A_1A_2}\frac{\partial A_2}{\partial \alpha_1}v\right).$$

Equations (I.125) now enable us to write the expanded expressions for the strain components of a shell in the curvilinear system α_1, α_2, z. It should be noted that the analogy obtaining between plates and shells allows us to make use of the same approximations in determining the deformations of shells as were used in the analysis of plates (§ 46).

The assumption that the straight fibers which are normal to the middle surface of the shell before the deformation remain straight and normal to the middle surface after deformation can be formulated analytically as follows:

$$\varepsilon_{1z}=\frac{\partial u}{\partial z}+\frac{1}{1+\frac{z}{R_1}}\left(\frac{1}{A_1}\frac{\partial w}{\partial\alpha_1}-\frac{u}{R_1}\right)+\frac{1}{1+\frac{z}{R_1}}\left(\frac{1}{A_1}\frac{\partial u}{\partial\alpha_1}\right.$$
$$\left.+\frac{1}{A_1A_2}\frac{\partial A_1}{\partial\alpha_2}v+\frac{w}{R_1}\right)\frac{\partial u}{\partial z}+\frac{1}{1+\frac{z}{R_2}}\left(\frac{1}{A_2}\frac{\partial u}{\partial\alpha_2}\right.$$
$$\left.-\frac{1}{A_1A_2}\frac{\partial A_2}{\partial\alpha_1}v\right)\frac{\partial v}{\partial z}+\frac{1}{1+\frac{z}{R_1}}\left(\frac{1}{A_1}\frac{\partial w}{\partial\alpha_1}-\frac{u}{R_1}\right)\frac{\partial w}{\partial z}=0,$$
$$\tag{VI.26}$$
$$\varepsilon_{2z}=\frac{\partial v}{\partial z}+\frac{1}{1+\frac{z}{R_2}}\left(\frac{1}{A_2}\frac{\partial w}{\partial\alpha_2}-\frac{v}{R_2}\right)+\frac{1}{1+\frac{z}{R_2}}\left(\frac{1}{A_2}\frac{\partial v}{\partial\alpha_2}\right.$$
$$\left.+\frac{1}{A_1A_2}\frac{\partial A_2}{\partial\alpha_1}u+\frac{w}{R_2}\right)\frac{\partial v}{\partial z}+\frac{1}{1+\frac{z}{R_2}}\left(\frac{1}{A_2}\frac{\partial u}{\partial\alpha_2}\right.$$
$$\left.-\frac{1}{A_1A_2}\frac{\partial A_2}{\partial\alpha_1}v\right)\frac{\partial u}{\partial z}+\frac{1}{1+\frac{z}{R_2}}\left(\frac{1}{A_2}\frac{\partial w}{\partial\alpha_2}-\frac{v}{R_2}\right)\frac{\partial w}{\partial z}=0.$$

The assumption that these normal fibers are not elongated leads to the equation

$$\varepsilon_{zz}=\frac{\partial w}{\partial z}+\frac{1}{2}\left[\left(\frac{\partial u}{\partial z}\right)^2+\left(\frac{\partial v}{\partial z}\right)^2+\left(\frac{\partial w}{\partial z}\right)^2\right]=0. \tag{VI.27}$$

We shall attempt to solve the system consisting of the three equations (VI.26), (VI.27) by taking the displacements u, v, w in the form

$$\begin{aligned} u&=\hat{u}(\alpha_1,\ \alpha_2)+z\vartheta(\alpha_1,\ \alpha_2),\\ v&=\hat{v}(\alpha_1,\ \alpha_2)+z\psi(\alpha_1,\ \alpha_2),\\ w&=\hat{w}(\alpha_1,\ \alpha_2)+z\chi(\alpha_1,\ \alpha_2). \end{aligned} \tag{VI.28}$$

Here, $\hat{u}$, $\hat{v}$, $\hat{w}$ are the displacements of the middle surface of the shell (which follows from equations (VI.28) by setting $z=0$).

If (VI.28) are substituted into (VI.27) and (VI.26), we get five equations, two of which, however, are implied by one of the other three. Hence, just three of the equations are independent, namely

$$\vartheta^2+\psi^2+(1+\chi)^2=1, \tag{VI.29}$$

$$(1+\chi)\left(\frac{1}{A_1}\frac{\partial \hat{w}}{\partial \alpha_1}-\frac{\hat{u}}{R_1}\right)+\left(\frac{1}{A_2}\frac{\partial \hat{u}}{\partial \alpha_2}-\frac{1}{A_1A_2}\frac{\partial A_2}{\partial \alpha_1}\hat{v}\right)\psi$$
$$+\left(1+\frac{1}{A_1}\frac{\partial \hat{u}}{\partial \alpha_1}+\frac{1}{A_1A_2}\frac{\partial A_1}{\partial \alpha_2}\hat{v}+\frac{\hat{w}}{R_1}\right)\vartheta=0,$$

$$(1+\chi)\left(\frac{1}{A_2}\frac{\partial \hat{w}}{\partial \alpha_2}-\frac{\hat{v}}{R_2}\right)+\left(\frac{1}{A_1}\frac{\partial \hat{v}}{\partial \alpha_1}-\frac{1}{A_1A_2}\frac{\partial A_1}{\partial \alpha_2}\hat{u}\right)\vartheta$$
$$+\left(1+\frac{1}{A_2}\frac{\partial \hat{v}}{\partial \alpha_2}+\frac{1}{A_1A_2}\frac{\partial A_2}{\partial \alpha_1}\hat{u}+\frac{\hat{w}}{R_2}\right)\psi=0.$$

On the basis of the last two of these three equations we have

$$\vartheta=\frac{\hat{\alpha}_{31}}{\hat{\alpha}_{23}}(1+\chi), \quad \psi=\frac{\hat{\alpha}_{32}}{\hat{\alpha}_{33}}(1+\chi), \tag{VI.30}$$

where, in this case,

$$\begin{aligned}\hat{\alpha}_{31}&=-\hat{e}_{13}(1+\hat{e}_{22})+\hat{e}_{23}\hat{e}_{21},\\ \hat{\alpha}_{32}&=-\hat{e}_{23}(1+\hat{e}_{11})+\hat{e}_{13}\hat{e}_{12},\\ \hat{\alpha}_{33}&=(1+\hat{e}_{11})(1+\hat{e}_{22})-\hat{e}_{12}\hat{e}_{21},\end{aligned} \tag{VI.31}$$

with

$$\begin{aligned}\hat{e}_{11}&=\frac{1}{A_1}\frac{\partial \hat{u}}{\partial \alpha_1}+\frac{1}{A_1A_2}\frac{\partial A_1}{\partial \alpha_2}\hat{v}+\frac{\hat{w}}{R_1},\\ \hat{e}_{22}&=\frac{1}{A_2}\frac{\partial \hat{v}}{\partial \alpha_2}+\frac{1}{A_1A_2}\frac{\partial A_2}{\partial \alpha_1}\hat{u}+\frac{\hat{w}}{R_2},\\ \hat{e}_{21}&=\frac{1}{A_1}\frac{\partial \hat{v}}{\partial \alpha_1}-\frac{1}{A_1A_2}\frac{\partial A_1}{\partial \alpha_2}\hat{u},\\ \hat{e}_{13}&=\frac{1}{A_1}\frac{\partial \hat{w}}{\partial \alpha_1}-\frac{\hat{u}}{R_1},\\ \hat{e}_{12}&=\frac{1}{A_2}\frac{\partial \hat{u}}{\partial \alpha_2}-\frac{1}{A_1A_2}\frac{\partial A_2}{\partial \alpha_1}\hat{v},\\ \hat{e}_{23}&=\frac{1}{A_2}\frac{\partial \hat{w}}{\partial \alpha_2}-\frac{\hat{v}}{R_2}.\end{aligned} \tag{VI.32}$$

Substituting (VI.30) into the first of equations (VI.29), we arrive at the following expression for χ:

$$\chi=\frac{\hat{\alpha}_{33}}{\sqrt{\hat{\alpha}_{31}^2+\hat{\alpha}_{32}^2+\hat{\alpha}_{33}^2}}-1, \tag{VI.33}$$

where (by considerations similar to those in § 46) we have the approximation

$$\hat{\alpha}_{31}^2+\hat{\alpha}_{32}^2+\hat{\alpha}_{33}^2\approx 1, \qquad \text{(VI.34)}$$

which is valid for elongations negligible compared to unity.

Hence

$$\chi\approx\hat{\alpha}_{33}-1=\hat{e}_{11}+\hat{e}_{22}+\hat{e}_{11}\hat{e}_{22}-\hat{e}_{12}\hat{e}_{21} \qquad \text{(VI.35)}$$

and substituting this value of χ into (VI.30), we obtain

$$\begin{aligned}\vartheta&=-\hat{e}_{13}(1+\hat{e}_{22})+\hat{e}_{23}\hat{e}_{12},\\ \psi&=-\hat{e}_{23}(1+\hat{e}_{11})+\hat{e}_{13}\hat{e}_{21}.\end{aligned} \qquad \text{(VI.36)}$$

Equations (VI.35), (VI.36), and (VI.28) express the displacements of an arbitrary point of the shell in terms of the displacements of the corresponding point of the middle surface. Since no approximations were made above, other than those which follow from the assumption that the elongations and shears are small, the result obtained is valid for arbitrary relative rotations of the shell, so long as they are consistent with the assumption that the strains are small. In other words, the derived equations are applicable to the strong bending of shells.

Further, let us determine the elongations and shears at a layer of the shell which is at a distance z from the middle surface. The corresponding strain components, on substitution of the Lamé coefficients (VI.22) into (I.125), become

$$\varepsilon_{11}=\frac{1}{1+\frac{z}{R_1}}\left(\frac{1}{A_1}\frac{\partial u}{\partial\alpha_1}+\frac{1}{A_1A_2}\frac{\partial A_1}{\partial\alpha_2}v+\frac{w}{R_1}\right)$$
$$+\frac{1}{2}\frac{1}{\left(1+\frac{z}{R_1}\right)^2}\left[\left(\frac{1}{A_1}\frac{\partial u}{\partial\alpha_1}+\frac{1}{A_1A_2}\frac{\partial A_1}{\partial\alpha_2}v+\frac{w}{R_1}\right)^2\right.$$

$$+\left(\frac{1}{A_1}\frac{\partial v}{\partial \alpha_1}-\frac{1}{A_1A_2}\frac{\partial A_1}{\partial \alpha_2}u\right)^2+\left(\frac{1}{A_1}\frac{\partial w}{\partial \alpha_1}-\frac{u}{R_1}\right)^2\Bigg],$$

$$\varepsilon_{22}=\frac{1}{1+\frac{z}{R_2}}\left(\frac{1}{A_2}\frac{\partial v}{\partial \alpha_2}+\frac{1}{A_1A_2}\frac{\partial A_2}{\partial \alpha_1}u+\frac{w}{R_2}\right)$$
$$+\frac{1}{2}\frac{1}{\left(1+\frac{z}{R_2}\right)^2}\Bigg[\left(\frac{1}{A_2}\frac{\partial u}{\partial \alpha_2}-\frac{1}{A_1A_2}\frac{\partial A_2}{\partial \alpha_1}v\right)^2$$
$$+\left(\frac{1}{A_2}\frac{\partial v}{\partial \alpha_2}+\frac{1}{A_1A_2}\frac{\partial A_2}{\partial \alpha_1}u+\frac{w}{R_2}\right)^2+\left(\frac{1}{A_2}\frac{\partial w}{\partial \alpha_2}-\frac{v}{R_2}\right)^2\Bigg], \quad \text{(VI.37)}$$

$$\varepsilon_{12}=\frac{1}{1+\frac{z}{R_1}}\left(\frac{1}{A_1}\frac{\partial v}{\partial \alpha_1}-\frac{1}{A_1A_2}\frac{\partial A_1}{\partial \alpha_2}u\right)+\frac{1}{1+\frac{z}{R_2}}\left(\frac{1}{A_2}\frac{\partial u}{\partial \alpha_2}\right.$$
$$\left.-\frac{1}{A_1A_2}\frac{\partial A_2}{\partial \alpha_1}v\right)+\frac{1}{\left(1+\frac{z}{R_1}\right)\left(1+\frac{z}{R_2}\right)}\Bigg[\left(\frac{1}{A_1}\frac{\partial u}{\partial \alpha_1}\right.$$
$$\left.+\frac{1}{A_1A_2}\frac{\partial A_1}{\partial \alpha_2}v+\frac{w}{R_1}\right)\left(\frac{1}{A_2}\frac{\partial u}{\partial \alpha_2}-\frac{1}{A_1A_2}\frac{\partial A_2}{\partial \alpha_1}v\right)$$
$$+\left(\frac{1}{A_1}\frac{\partial v}{\partial \alpha_1}-\frac{1}{A_1A_2}\frac{\partial A_1}{\partial \alpha_2}u\right)\left(\frac{1}{A_2}\frac{\partial v}{\partial \alpha_2}+\frac{1}{A_1A_2}\frac{\partial A_2}{\partial \alpha_1}u\right.$$
$$\left.+\frac{w}{R_2}\right)+\left(\frac{1}{A_1}\frac{\partial w}{\partial \alpha_1}-\frac{u}{R_1}\right)\left(\frac{1}{A_2}\frac{\partial w}{\partial \alpha_2}-\frac{v}{R_2}\right)\Bigg].$$

Since $-\delta/_2 \leqslant z \leqslant \delta/_2$, in the case of thin shells one can set

$$1+\frac{z}{R_1}\approx 1+\frac{z}{R_2}\approx 1 \quad \text{(VI.38)}$$

in the above equations. Making use of this approximation and then substituting the displacements u, v, w in the form (VI.28) into (VI.37), we obtain

$$\begin{aligned}\varepsilon_{11}&=\hat{\varepsilon}_{11}+z\varkappa_{11}+z^2\nu_{11},\\ \varepsilon_{22}&=\hat{\varepsilon}_{22}+z\varkappa_{22}+z^2\nu_{22},\\ \varepsilon_{12}&=\hat{\varepsilon}_{12}+z\varkappa_{12}+z^2\nu_{12}.\end{aligned} \quad \text{(VI.39)}$$

Here $\hat{\varepsilon}_{11}$, $\hat{\varepsilon}_{22}$, $\hat{\varepsilon}_{12}$ are the elongations and shear of the middle surface of the shell. These are given by

$$\hat{\varepsilon}_{11} = \hat{e}_{11} + {}^1/_2\,[\hat{e}_{11}^2 + \hat{e}_{12}^2 + \hat{e}_{13}^2],$$
$$\hat{\varepsilon}_{22} = \hat{e}_{22} + {}^1/_2\,[\hat{e}_{21}^2 + \hat{e}_{22}^2 + \hat{e}_{23}^2], \qquad \text{(VI.40)}$$
$$\hat{\varepsilon}_{12} = \hat{e}_{12} + \hat{e}_{21} + \hat{e}_{11}\hat{e}_{21} + \hat{e}_{22}\hat{e}_{12} + \hat{e}_{13}\hat{e}_{23}.$$

The parameters $\varkappa_{11}, \varkappa_{22}, \varkappa_{12}$, which characterize the variation of the curvature of the middle surface induced by the deformation, are in turn given by

$$\varkappa_{11} = (1 + \hat{e}_{11})\,k_{11} + \hat{e}_{12}k_{12} + \hat{e}_{13}k_{13},$$
$$\varkappa_{22} = (1 + \hat{e}_{22})\,k_{22} + \hat{e}_{21}k_{21} + \hat{e}_{23}k_{23}, \qquad \text{(VI.41)}$$
$$\varkappa_{12} = k_{21}(1 + \hat{e}_{11}) + k_{12}(1 + \hat{e}_{22}) + k_{11}\hat{e}_{21} + k_{22}\hat{e}_{12} + k_{13}\hat{e}_{23} + k_{23}\hat{e}_{13},$$

where

$$k_{11} = \frac{1}{A_1}\frac{\partial\vartheta}{\partial\alpha_1} + \frac{1}{A_1A_2}\frac{\partial A_1}{\partial\alpha_2}\psi + \frac{\chi}{R_1},$$
$$k_{22} = \frac{1}{A_2}\frac{\partial\psi}{\partial\alpha_2} + \frac{1}{A_1A_2}\frac{\partial A_2}{\partial\alpha_1}\vartheta + \frac{\chi}{R_2}, \qquad \text{(VI.42)}$$
$$k_{21} = \frac{1}{A_2}\frac{\partial\vartheta}{\partial\alpha_2} - \frac{1}{A_1A_2}\frac{\partial A_2}{\partial\alpha_1}\psi, \quad k_{12} = \frac{1}{A_1}\frac{\partial\psi}{\partial\alpha_1} - \frac{1}{A_1A_2}\frac{\partial A_1}{\partial\alpha_2}\vartheta,$$
$$k_{13} = \frac{1}{A_1}\frac{\partial\chi}{\partial\alpha_1} - \frac{\vartheta}{R_1}, \quad k_{23} = \frac{1}{A_2}\frac{\partial\chi}{\partial\alpha_2} - \frac{\psi}{R_2}.$$

There is no need to give the expressions for $\nu_{11}, \nu_{22}, \nu_{12}$, since the corresponding terms in equations (VI.39) are always negligible for small elongations and shears (just as in the case of plates).

The above theory exhausts the analysis of the deformation of thin shells for small strains and arbitrary rotations. If in the above equations one retains only terms which are linear in the displacements and their derivatives, one arrives at the ordinary theory of deformation of thin shells. Similarly, one may derive from the above equations a theory of deformation of shells analogous in its accuracy to T. v. Kármán's theory of plates.

§ 49. On the Nature of Kirchhoff's Assumptions

Let us now clarify in greater detail the mathematical foundation of the assumptions adopted above in the theory of the deformation of plates and shells.

To do so, we shall seek an exact solution of the problem of the deformation of a plate by expanding the displacement components into power series:

$$u = u\,(x,\ y,\ o) + \left(\frac{\partial u}{\partial z}\right)_0 z + \frac{1}{2}\left(\frac{\partial^2 u}{\partial z^2}\right)_0 z^2 + \ldots,$$
$$v = v\,(x,\ y,\ o) + \left(\frac{\partial v}{\partial z}\right)_0 z + \frac{1}{2}\left(\frac{\partial^2 v}{\partial z^2}\right)_0 z^2 + \ldots, \qquad \text{(VI.43)}$$
$$w = w\,(x,\ y,\ o) + \left(\frac{\partial w}{\partial z}\right)_0 z + \frac{1}{2}\left(\frac{\partial^2 w}{\partial z^2}\right)_0 z^2 + \ldots.$$

Retaining in these series only the first two terms and introducing the notations

$$u\,(x,\ y,\ o) = \hat{u}, \quad v\,(x,\ y,\ o) = \hat{v}, \quad w\,(x,\ y,\ o) = \hat{w}$$
$$\left(\frac{\partial u}{\partial z}\right)_0 = \vartheta, \quad \left(\frac{\partial v}{\partial z}\right)_0 = \psi, \quad \left(\frac{\partial w}{\partial z}\right)_0 = \chi, \qquad \text{(VI.44)}$$

we obtain (VI.2).

Thus, we may say that in treating the deformation of a thin plate, Kirchhoff's assumptions, first of all, assert that it suffices to retain only the first two terms in the series (VI.43). However, these assumptions establish, in addition, a relation between the displacements $\hat{u}$, $\hat{v}$, $\hat{w}$ of the middle surface of the plate and the parameters ϑ, ψ, χ , allowing us to express the second set of parameters in terms of the first.

In order to clarify the nature of this relation, we recall that in accordance with Table I of § 2

$$\frac{1}{\sqrt{1+2\varepsilon_{zz}}}\frac{\partial u}{\partial z}=\cos(X,\ \mathbf{i}_3),\quad \frac{1}{\sqrt{1+2\varepsilon_{zz}}}\frac{\partial v}{\partial z}=\cos(Y,\ \mathbf{i}_3),$$
$$\frac{1}{\sqrt{1+2\varepsilon_{zz}}}\left(1+\frac{\partial w}{\partial z}\right)=\cos(Z,\ \mathbf{i}_3), \qquad \text{(VI.45)}$$

where $\mathbf{i}_3$ is a unit vector tangent to a fiber of the plate which was straight and perpendicular to the middle surface before the deformation.

It follows from (VI.44) and (VI.45) that

$$\vartheta=\sqrt{1+2\hat{\varepsilon}_{zz}}\cos(X,\ \hat{\mathbf{i}}_3),\quad \psi=\sqrt{1+2\hat{\varepsilon}_{zz}}\cos(Y,\ \hat{\mathbf{i}}_3),$$
$$1+\chi=\sqrt{1+2\hat{\varepsilon}_{zz}}\cos(Z,\ \hat{\mathbf{i}}_3), \qquad \text{(VI.46)}$$

where $\hat{\mathbf{i}}_3$ is in the direction of the tangent to the same fiber at its point of intersection with the middle surface and $\hat{\varepsilon}_{zz}$ is the elongation of the fiber at that point.

Equations (VI.46) imply that if the elongations are neglected in comparison with unity, the parameters ϑ, ψ, and $1+\chi$ can be identified with the direction cosines of that fiber of the plate in the strained state which in the unstrained state was normal to the middle surface. In the problem of bending of plates, however, the angles of rotation are always large compared to the shears.

To determine the unit vector $\hat{\mathbf{i}}_3$, i. e., to determine the parameters ϑ, ψ, χ, one may neglect those shears which affect the direction of $\hat{\mathbf{i}}_3$.

In other words, in determining ϑ, ψ, χ, one may use the equations

$$\varepsilon_{xy}\approx\varepsilon_{yz}\approx 0. \qquad \text{(VI.47)}$$

Furthermore, if $\hat{\varepsilon}_{zz}$ is neglected compared to unity, then, squaring each of the three equations and adding the results we obtain

$$\vartheta^2+\psi^2+(1+\chi)^2=1.$$

This, as is known from § 46, is equivalent to

$$\hat{\varepsilon}_{zz}=0. \tag{VI.48}$$

This equation, therefore, is a statement of the fact that the elongation of a fiber normal to the middle surface of the plate has a negligible effect on its direction after the deformation.

The above argument can be divided into two parts.

In the first part it was shown how the expansions of the displacements into power series in z, where z is the distance of a point of the plate from the middle surface, can be treated on the basis of the fundamental assumptions of the theory of plates. This enabled us to determine the relation between ϑ, ψ, χ and the derivatives of the displacements u, v, w with respect to z. It turned out that ϑ, ψ, and χ characterize the direction in the strained state of a fiber of the plate which is initially normal to the middle surface.

In the second part the strains which affect the direction of this fiber were neglected in comparison with the rotations, which led to the equations (VI.47) and (VI.48).

In § 46, conversely, we postulated equations (VI.47), (VI.48), from which all the rest followed immediately. This way is shorter, but less intuitive, since equations (VI.47) and (VI 48) give no idea of what is being neglected or what basis of comparison is used.

Interpreted as mathematically exact relations, (VI.47) and (VI.48) are absurd, since they lead, in general, to contradictions in the formulation of the conditions of equilibrium of an element of the plate. However, if they are interpreted as meaning only that certain strain components are neglected in determining $\hat{i}_3$, equations (VI.47) and (VI.48) are seen in the right perspective. At the same time, those conditions under which these equations can be used are also immediately evident.

This analysis for plates obviously also applies to shells.

The following conclusions may be inferred from the above:

a) Kirchhoff's hypothesis in the theory of plates and shells rests on simplifications which result when elongations and shears are neglected in comparison with rotations in determining the direction of fibers of the strained body.

Since thin plates and shells are flexible bodies whose angles of rotation under a deformation ordinarily are large in comparison with the elongations and shears, the adoption of this hypothesis usually introduces only a negligible error into the calculations. Hence, it is clear that the simplification in the theory of plates proposed by Kirchhoff, and subsequently extended by Love to shells, can hardly be called a hypothesis as is ordinarily done. For, in essence, we are dealing with purely geometric approximations whose error can always be estimated.

b) Kirchhoff's "hypothesis" does not include any assumptions about the properties of the materials of which the plates and shells are made. Thus, a theory based on this "hypothesis" can be used with equal effectiveness both for bodies which obey Hooke's law and for bodies which do not. It is only important that the basic condition be satisfied, namely, that the strains be small in comparison with the angles of rotation.

This remark emphasizes still more the non-hypothetical character of the approximations adopted in the theory of plates and shells.

When the "hypothesis" of plane sections is applied to beams which are subjected to deformations in the elastic-plastic domain, it is generally considered necessary to refer to an experiment which confirms the "hypothesis" in this case. It is clear from the above that the designing of special experiments to verify the "hypothesis" of plane sections for bending in the elastic-plastic domain must be ascribed to an insufficiently clear understanding of the nature of this hypothesis.

§ 50. Deformation of Rods (First Approximation)

Having developed the theory of thin plates and shells, we shall now make use of analogous methods in the theory of thin rods. As will become apparent, this problem is somewhat more complicated than the two preceding ones. This is the reason for considering it after them.

Let a thin prismatic rod of arbitrary cross section be given (Fig. 20).

We place the origin of the system X, Y, Z at the center of gravity of the area of one of the ends of the rod, and direct the Z-axis along the rod, and the X- and Y-axes along the principal axes of inertia of the cross-section. We note that in the present chapter we could have directed the axes differently. Accordingly, it would not have been necessary to place the origin of the Cartesian system specifically at the center of gravity of an end of the rod. However, since the above choice of coordinates has definite advantages in studying the conditions of equilibrium of a volume element of the rod, we shall adhere to this choice in the present section.

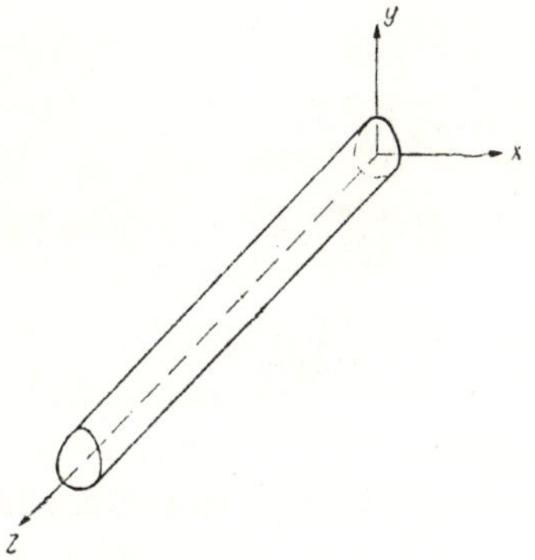

Fig. 20

Denote by $u\ (x, y, z)$, $v\ (x, y, z)$, $w\ (x, y, z)$ the displacements of an arbitrary point of the rod due to a deformation. Since the variations of the x and y coordinates in this problem are substantially smaller than the variation of the z coordinate, we may assume that the power series expansions of the displacements in x and y converge rapidly enough within limits which are of interest to us.

We shall base the theory of rods on this consideration, whose value was shown in the theory of plates and shells.

Accordingly, the displacements of an arbitrary point of the bar can be expressed in the form

$$\begin{aligned} u\ (x, y, z) &= \hat{u}\ (z) + x\vartheta_1\ (z) \\ &\quad + y\vartheta_2\ (z) + \bar{u}\ (x, y, z), \\ v\ (x, y, z) &= \hat{v}\ (z) + x\psi_1\ (z) \\ &\quad + y\psi_2\ (z) + \bar{v}\ (x, y, z), \\ w\ (x, y, z) &= \hat{w}\ (z) + x\chi_1\ (z) \\ &\quad + y\chi_2\ (z) + \bar{w}\ (x, y, z), \end{aligned} \tag{VI.49}$$

where

$$\hat{u} = u\ (o, o, z),\quad \hat{v} = v\ (o, o, z),\quad \hat{w} = w\ (o, o, z) \tag{VI.50}$$

$$\vartheta_1 = \left(\frac{\partial u}{\partial x}\right) \quad = \left(\frac{\partial u}{\partial x}\right)_0,\quad \psi_1 = \left(\frac{\partial v}{\partial x}\right)_0,\quad \chi_1 = \left(\frac{\partial w}{\partial x}\right)_0,$$

$$\text{for } x=0, \quad y=0 \tag{VI.51}$$

$$\vartheta_2=\left(\frac{\partial u}{\partial y}\right)=\left(\frac{\partial u}{\partial y}\right)_0, \quad \psi_2=\left(\frac{\partial v}{\partial y}\right)_0, \quad \chi_2=\left(\frac{\partial w}{\partial y}\right)_0,$$

$$\text{for } x=0, \quad y=0$$

$$\bar{u}=\frac{1}{2}x^2\left(\frac{\partial^2 u}{\partial x^2}\right)_0+\frac{1}{2}y^2\left(\frac{\partial^2 u}{\partial y^2}\right)_0+xy\left(\frac{\partial^2 u}{\partial x\partial y}\right)_0+\ldots,$$

$$\bar{v}=\frac{1}{2}x^2\left(\frac{\partial^2 v}{\partial x^2}\right)_0+\frac{1}{2}y^2\left(\frac{\partial^2 v}{\partial y^2}\right)_0+xy\left(\frac{\partial^2 v}{\partial x\partial y}\right)_0+\ldots, \tag{VI.52}$$

$$\bar{w}=\frac{1}{2}x^2\left(\frac{\partial^2 w}{\partial x^2}\right)_0+\frac{1}{2}y^2\left(\frac{\partial^2 w}{\partial y^2}\right)_0+xy\left(\frac{\partial^2 w}{\partial x\partial y}\right)_0+\ldots.$$

The last equations imply that

a) $\hat{u}$, $\hat{v}$, $\hat{w}$ are the displacements of the points on the axis of the rod.

b) $1+\vartheta_1$, ϑ_2, ψ_1, $1+\psi_2$, χ_1, χ_2 are of the same order of magnitude as the direction cosines of those fibers in the strained state which were initially parallel to the X- and Y-axes (Table 1, § 2). We shall assume in the sequel that these parameters (some or all of them) substantially exceed the elongations and shears.

c) $\bar{u}$, $\bar{v}$, $\bar{w}$ contain all terms, beginning with the fourth, of the power series for the displacements.

It is clear from this that for $x=0$, $y=0$,

$$\bar{u}=\bar{v}=\bar{w}=\frac{\partial \bar{u}}{\partial x}=\frac{\partial \bar{u}}{\partial y}=\frac{\partial \bar{v}}{\partial x}=\frac{\partial \bar{v}}{\partial y}=\frac{\partial \bar{w}}{\partial x}=\frac{\partial \bar{w}}{\partial y}=0. \tag{VI.53}$$

In the sequel $\bar{u}$, $\bar{v}$, $\bar{w}$ will be regarded as the correction terms in equations (VI.49), which are very small in comparison with the remaining terms. If $\bar{u}$, $\bar{v}$, $\bar{w}$ are neglected, the equations (VI.49) become

$$\begin{aligned} u&=\hat{u}+x\vartheta_1+y\vartheta_2,\\ v&=\hat{v}+x\psi_1+y\psi_2,\\ w&=\hat{w}+x\chi_1+y\chi_2. \end{aligned} \tag{VI.54}$$

It is evident that taking the displacements of the rod in this form corresponds to that degree of accuracy which was adopted above in the theory of plates and shells.

The difference between (VI.2) and (VI.54) is that, in the first case, the series are expanded in terms of the one coordinate z, while, in the second case, the expansion is in terms of the two coordinates x and y. This is true because for plates only the z coordinate varies between narrow limits, while for rods the two coordinates x and y possess this property. Equations (VI.2) and (VI.54) are entirely analogous as far as the retention of terms is concerned, since the series are ended in both cases at the terms which depend linearly on those coordinates with respect to which they are expanded.

While the accuracy of (VI.2) is entirely adequate for plates and shells, the same is not true in the case of equations (VI.54) for rods, since a solution in this form cannot be subjected to boundary conditions which arise in practice. Hence, in studying the deformation of rods, it becomes necessary to take the displacements in the more complicated form (VI.49), rather than in the form (VI.54). This constitutes the main difference between the theory of rods and the theory of plates and shells. However, for greater clarity, we shall begin by assuming that equations (VI.54) are adequate. After completing the computations, we shall find that we have not retained enough terms to give a full solution of the problem. At this point we shall introduce the necessary corrections. This will cause no special difficulties, since by that time the reader will have a complete picture of the method.

Substituting equations (VI.54) into the expressions (I.22) for the strain components, we obtain

$$\begin{aligned}
&\varepsilon_{xx}=\hat{\varepsilon}_{xx},\quad \varepsilon_{yy}=\hat{\varepsilon}_{yy},\quad \varepsilon_{xy}=\hat{\varepsilon}_{xy},\\
&\varepsilon_{xz}=\hat{\varepsilon}_{xz}+x\frac{d\hat{\varepsilon}_{xx}}{dz}+yk_{xy},\\
&\varepsilon_{yz}=\hat{\varepsilon}_{yz}+y\frac{d\hat{\varepsilon}_{yy}}{dz}+x\left(\frac{d\hat{\varepsilon}_{xy}}{dz}-k_{xy}\right),\qquad\text{(VI.55)}\\
&\varepsilon_{zz}=\hat{\varepsilon}_{zz}+xk_{xx}+yk_{yy}+x^2\nu_{xx}+y^2\nu_{yy}+xy\nu_{xy},
\end{aligned}$$

where

$$\begin{aligned}
&\hat{\varepsilon}_{xx}=\vartheta_1+\frac{1}{2}(\vartheta_1^2+\psi_1^2+\chi_1^2),\\
&\hat{\varepsilon}_{yy}=\psi_2+\frac{1}{2}(\vartheta_2^2+\psi_2^2+\chi_2^2),\\
&\hat{\varepsilon}_{xy}=\vartheta_2(1+\vartheta_1)+\psi_1(1+\psi_2)+\chi_1\chi_2,\qquad\text{(VI.56)}\\
&\hat{\varepsilon}_{xz}=(1+\vartheta_1)\frac{d\hat{u}}{dz}+\psi_1\frac{d\hat{v}}{dz}+\chi_1\left(1+\frac{d\hat{w}}{dz}\right),\\
&\hat{\varepsilon}_{yz}=\vartheta_2\frac{d\hat{u}}{dz}+(1+\psi_2)\frac{d\hat{v}}{dz}+\chi_2\left(1+\frac{d\hat{w}}{dz}\right),\\
&\hat{\varepsilon}_{zz}=\frac{d\hat{w}}{dz}+\frac{1}{2}\left[\left(\frac{d\hat{u}}{dz}\right)^2+\left(\frac{d\hat{v}}{dz}\right)^2+\left(\frac{d\hat{w}}{dz}\right)^2\right]
\end{aligned}$$

are the values of the strain components on the axis of the rod, and consequently are functions of z alone.

The quantities k_{xx}, k_{yy}, k_{xy} are also functions of z alone and are given by

$$\begin{aligned}
&k_{xx}=\frac{d\hat{u}}{dz}\frac{d\vartheta_1}{dz}+\frac{d\hat{v}}{dz}\frac{d\psi_1}{dz}+\left(1+\frac{d\hat{w}}{dz}\right)\frac{d\chi_1}{dz},\\
&k_{yy}=\frac{d\hat{u}}{dz}\frac{d\vartheta_2}{dz}+\frac{d\hat{v}}{dz}\frac{d\psi_2}{dz}+\left(1+\frac{d\hat{w}}{dz}\right)\frac{d\chi_2}{dz},\qquad\text{(VI.57)}\\
&k_{xy}=(1+\vartheta_1)\frac{d\vartheta_2}{dz}+\psi_1\frac{d\psi_2}{dz}+\chi_1\frac{d\chi_2}{dz}.
\end{aligned}$$

Finally, the coefficients ν_{xx}, ν_{yy}, ν_{xy} are

$$\nu_{xx}=\frac{1}{2}\left[\left(\frac{d\vartheta_1}{dz}\right)^2+\left(\frac{d\psi_1}{dz}\right)^2+\left(\frac{d\chi_1}{dz}\right)^2\right],$$

$$\nu_{yy}=\frac{1}{2}\left[\left(\frac{d\vartheta_2}{dz}\right)^2+\left(\frac{d\psi_2}{dz}\right)^2+\left(\frac{d\chi_2}{dz}\right)^2\right], \qquad \text{(VI.58)}$$

$$\nu_{xy}=\frac{d\vartheta_1}{dz}\frac{d\vartheta_2}{dz}+\frac{d\psi_1}{dz}\frac{d\psi_2}{dz}+\frac{d\chi_1}{dz}\frac{d\chi_2}{dz}.$$

As was pointed out before, all the parameters $\vartheta_1, \vartheta_2, \psi_1, \psi_2, \chi_1, \chi_2$, or at least some of them, must be regarded as substantially exceeding the strain components, since in the bending of a thin rod some, or all, of the angles of rotation are large in comparison with the elongations and shears. For the same reason, the derivatives $\frac{d\hat{u}}{dz}, \frac{d\hat{v}}{dz}, \frac{d\hat{w}}{dz}$ (Table 1, § 2) possess the same property. Hence it follows that the right-hand sides of (VI.56) must represent small differences of large terms. Consequently, in determining the nine parameters listed above, we may use the six equations

$$\hat{\varepsilon}_{xx}=\hat{\varepsilon}_{yy}=\hat{\varepsilon}_{xy}=\hat{\varepsilon}_{xz}=\hat{\varepsilon}_{yz}=\hat{\varepsilon}_{zz}=0. \qquad \text{(VI.59)}$$

Here, naturally, the equations should not be interpreted as meaning that all the strain components of the rod along its axis are negligible. Rather, they must be understood in the sense pointed out above.

The six equations (VI.59) are satisfied identically by setting

$$\vartheta_1=\cos\alpha\cos\beta\cos\gamma-\sin\alpha\sin\beta-1, \quad \vartheta_2=\cos\beta\sin\gamma,$$

$$\psi_1=-\cos\alpha\sin\gamma, \quad \psi_2=\cos\gamma-1, \qquad \text{(VI.60)}$$

$$\chi_1=\cos\alpha\sin\beta\cos\gamma+\sin\alpha\cos\beta, \quad \chi_2=\sin\beta\sin\gamma,$$

$$\frac{d\hat{u}}{dz}=-\cos\alpha\sin\beta-\sin\alpha\cos\beta\cos\gamma, \qquad \text{(VI.61)}$$

$$\frac{d\hat{v}}{dz}=\sin\alpha\sin\gamma, \quad \frac{d\hat{w}}{dz}=\cos\alpha\cos\beta-\sin\alpha\sin\beta\cos\gamma-1.$$

Here α, β, γ are three arbitrary angles which are functions of z alone (since $\hat{u}$, $\hat{v}$, $\hat{w}$ are functions of that coordinate alone).

The geometric meaning of equations (VI.60) and (VI.61) is easy to see. To this end, let us examine an arbitrary point M on the axis of the rod and the three line elements ax, ay, dz emanating from it (Fig. 21).

As a result of the deformation, the point M is displaced by the amounts $\hat{u}$, $\hat{v}$, $\hat{w}$ and assumes the position M^*, while the line elements are

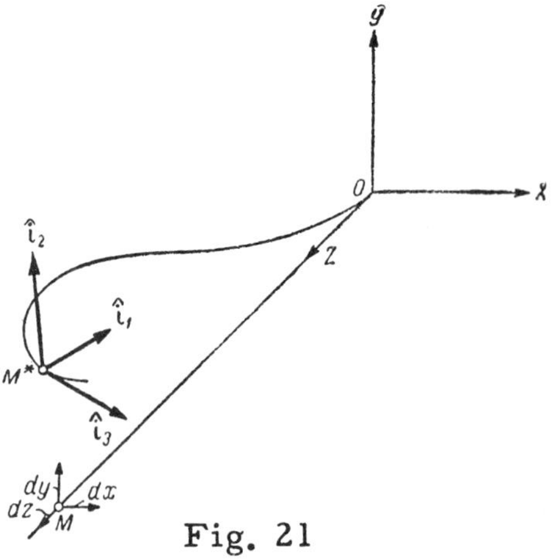

Fig. 21

directed along $\mathbf{i}_1$, $\mathbf{i}_2$, $\mathbf{i}_3$. If the angles of rotation of the elements of the rod are large in comparison with the shears, the latter may be neglected in determining the directions $\mathbf{i}_1$, $\mathbf{i}_2$, $\mathbf{i}_3$. With this approximation, the trihedral $\mathbf{i}_1$, $\mathbf{i}_2$, $\mathbf{i}_3$ can be taken as orthogonal and the parameters

$$1+\vartheta_1, \quad \vartheta_2, \quad \psi_1, \quad 1+\psi_2, \quad \chi_1, \quad \chi_2, \quad \frac{d\hat{u}}{dz}, \ \frac{d\hat{v}}{dz},$$

$$1+\frac{d\hat{w}}{dz}$$

become equal to the direction cosines of $\mathbf{i}_1$, $\mathbf{i}_2$, $\mathbf{i}_3$ if the elongations are neglected in comparison with unity. The latter becomes clear from equations (VI.54) and Tables 1 and 5 of Chapter I.

However, as is well-known, a system of three mutually perpendicular directions can be fixed with respect to another such system by means of the three Euler angles. Introducing these angles in accordance with Fig. 22, the above leads to equations (VI.60) and (VI.61).

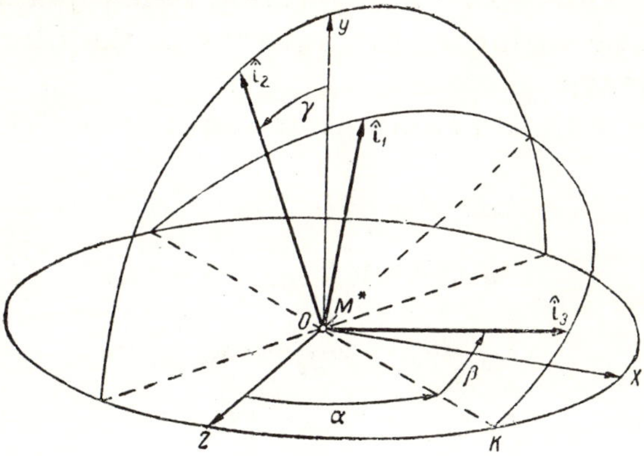

Fig. 22

Thus, the angles α, β, γ in equations (VI.60), (VI.61) are the Euler angles which determine the directions $\mathbf{i}_1$, $\mathbf{i}_2$, $\mathbf{i}_3$ with respect to X, Y, Z (to within the angles of shear).

If (VI.60) and (VI.61) are substituted in (VI.58), we get

$$
\begin{aligned}
k_{xx} &= \frac{d\alpha}{dz} + \frac{d\beta}{dz}\cos\gamma,\\
k_{yy} &= \cos\alpha\sin\gamma\frac{d\beta}{dz} - \sin\alpha\frac{d\gamma}{dz},\\
k_{xy} &= \cos\alpha\frac{d\gamma}{dz} + \sin\alpha\sin\gamma\frac{d\beta}{dz},
\end{aligned}
\qquad \text{(VI.62)}
$$

and

$$\nu_{xx} = \frac{1}{2}(k_{xx}^2 + k_{xy}^2),\ \nu_{yy} = \frac{1}{2}(k_{yy}^2 + k_{xy}^2),\ \nu_{xy} = k_{xx}k_{yy}. \qquad \text{(VI.63)}$$

If these values of the coefficients $\nu_{xx}, \nu_{yy}, \nu_{xy}$ are substituted into the last of equations (VI.55), the result is

$$\varepsilon_{zz} = \hat{\varepsilon}_{zz} + xk_{xx} + yk_{yy} + \frac{1}{2}\left(k_{xx}^2 + k_{xy}^2\right)x^2 + \frac{1}{2}\left(k_{yy}^2 + k_{xy}^2\right)y^2 + k_{xx}k_{yy}xy. \tag{VI.64}$$

It is seen that the terms corresponding to these coefficients can be neglected, being quantities of the same order as the squares of the elongations and shears.

With this approximation, equations (VI.55) become

$$\begin{gathered} \varepsilon_{xx} = \hat{\varepsilon}_{xx}, \quad \varepsilon_{yy} = \hat{\varepsilon}_{yy}, \quad \varepsilon_{xy} = \hat{\varepsilon}_{xy} \\ \varepsilon_{xz} = \hat{\varepsilon}_{xz} + x\frac{d\hat{\varepsilon}_{xx}}{dz} + yk_{xy} \\ \varepsilon_{yz} = \hat{\varepsilon}_{yz} + y\frac{d\hat{\varepsilon}_{yy}}{dz} + x\left(\frac{d\hat{\varepsilon}_{xy}}{dz} - k_{xy}\right) \\ \varepsilon_{zz} = \hat{\varepsilon}_{zz} + xk_{xx} + yk_{yy}, \end{gathered} \tag{VI.65}$$

where k_{xx}, k_{yy}, k_{xy} are determined by (VI.62).

Since the derivatives $\frac{d\hat{\varepsilon}_{xx}}{dz}, \frac{d\hat{\varepsilon}_{yy}}{dz}, \frac{d\hat{\varepsilon}_{xy}}{dz}$ are ordinarily small compared to k_{xx}, k_{yy}, k_{xy}, which characterize the curvature of the axis of the rod in the strained state, the terms in equations (VI.65) containing these derivatives may be neglected. The result is

$$\begin{gathered} \varepsilon_{xx} = \hat{\varepsilon}_{xx}, \quad \varepsilon_{yy} = \hat{\varepsilon}_{yy}, \quad \varepsilon_{xy} = \hat{\varepsilon}_{xy}, \\ \varepsilon_{xz} = \hat{\varepsilon}_{xz} + yk_{xy}, \quad \varepsilon_{yz} = \hat{\varepsilon}_{yz} - xk_{xy}, \quad \varepsilon_{zz} = \hat{\varepsilon}_{zz} + xk_{xx} + yk_{yy}. \end{gathered} \tag{VI.66}$$

These equations are based on the assumption that the elongations and shears are negligibly small in comparison with unity and the angles of rotation of the elements of the rod. However, in deriving equations (VI.66) it was postulated that only the first three terms of the Taylor series for the displacements need be retained. This leads to equations (VI.54). This assumption is not obvious, and moreover is not even correct, as is seen by applying equations (VI.66) to the special case in which the rod is not bent but only twisted uniformly along its whole length.

In that case

$$\alpha=\beta=0,\ \gamma=\tau z, \tag{VI.67}$$

where τ is a constant coefficient.

Substituting these values of the Euler angles into (VI.60), (VI.61), we find

$$\begin{aligned}&\vartheta_1=-(1-\cos\tau z),\ \vartheta_2=\sin\tau z,\\ &\psi_1=-\sin\tau z,\ \psi_2=-(1-\cos\tau z),\\ &\chi_1=\chi_2=\frac{d\hat{u}}{dz}=\frac{d\hat{v}}{dz}=\frac{d\hat{w}}{dz}=0.\end{aligned} \tag{VI.68}$$

Hence we obtain the following equations for the displacements:

$$\begin{aligned}&u=-x(1-\cos\tau z)-y\sin\tau z,\\ &v=+x\sin\tau z-y(1-\cos\tau z),\\ &w=0\end{aligned} \tag{VI.69}$$

and for the strain components we get

$$\begin{aligned}&\varepsilon_{xx}=\varepsilon_{yy}=\varepsilon_{xy}=\varepsilon_{zz}=0,\\ &\varepsilon_{xz}=y\tau \qquad \varepsilon_{yz}=-x\tau.\end{aligned} \tag{VI.70}$$

These expressions coincide with the "old" theory of torsion, rather than with the Saint-Venant theory. The former, as is well-known, is inadequate since it does not permit the freeing of the lateral surface of the rod from stresses, which is essential in this problem. Hence it is clear that the general equations (VI.66) are also inadequate and must be corrected so as to yield Saint-Venant's theory of torsion as a special case. In order to correct the results of this section we must abandon equations (VI.54) and return to equations (VI.49) in which $\bar{u}$, $\bar{v}$, $\bar{w}$ represent the remaining terms of the power series.

§ 51. Deformation of Rods (Second Approximation)

Starting this time with equations (VI.54), instead of (VI.49), we repeat the preceding calculations.

We shall consider the supplementary displacements $\bar{u}$, $\bar{v}$, $\bar{w}$, which up to now have not been taken into account, to be small and neglect terms which contain their products or products of their derivatives.

Substituting (VI.49) into the equations (I.22), we obtain

$$\varepsilon_{xx}=\hat{\varepsilon}_{xx}+\frac{\partial \mathfrak{u}}{\partial x},\quad \varepsilon_{yy}=\varepsilon_{yy}+\frac{\partial \mathfrak{v}}{\partial y},\quad \varepsilon_{xy}=\hat{\varepsilon}_{xy}+\frac{\partial \mathfrak{u}}{\partial y}+\frac{\partial \mathfrak{v}}{\partial x},$$

$$\varepsilon_{xz}=\hat{\varepsilon}_{xz}+x\frac{d\hat{\varepsilon}_{xx}}{dz}+yk_{xy}+\frac{\partial \mathfrak{u}}{\partial z}+\frac{\partial \mathfrak{w}}{\partial x}$$
$$\underline{\underline{-2\left(\bar{u}\frac{d\vartheta_1}{dz}+\bar{v}\frac{d\psi_1}{dz}+\bar{w}\frac{d\chi_1}{dz}\right)}},$$

$$\varepsilon_{yz}=\hat{\varepsilon}_{yz}+y\frac{d\hat{\varepsilon}_{yy}}{dz}+x\left(\frac{d\hat{\varepsilon}_{xy}}{dz}-k_{xy}\right)+\frac{\partial \mathfrak{v}}{\partial z}+\frac{\partial \mathfrak{w}}{\partial y}$$
$$\underline{\underline{-2\left(\bar{u}\frac{d\vartheta_2}{dz}+\bar{v}\frac{d\psi_2}{dz}+\bar{w}\frac{d\chi_2}{dz}\right)}},$$

$$\varepsilon_{zz}=\hat{\varepsilon}_{zz}+xk_{xx}+yk_{yy}+x^2\nu_{xx}+y^2\nu_{yy}+xy\nu_{xy}+\frac{\partial \mathfrak{w}}{\partial z}$$
$$\underline{\underline{-\bar{u}\left(\frac{d^2\hat{u}}{dz^2}+x\frac{d^2\vartheta_1}{dz^2}+y\frac{d^2\vartheta_2}{dz^2}\right)-\bar{v}\left(\frac{d^2\hat{v}}{dz^2}+x\frac{d^2\psi_1}{dz^2}\right.}} \qquad \text{(VI.71)}$$
$$\underline{\underline{\left.+y\frac{d^2\psi_2}{dz^2}\right)-\bar{w}\left(\frac{d^2w}{dz^2}+x\frac{d^2\chi_1}{dz^2}+y\frac{d^2\chi_2}{dz^2}\right)}}.$$

As before, $\varepsilon_{xx}, \ldots, \varepsilon_{zz}$ are determined by equations (VI.55), $k_{xx}, k_{yy}, k_{xy}, \nu_{xx}, \nu_{yy}, \nu_{xy}$ by equations (VI.57) and (VI.58), and u, v, w by (VI.52).

The new letters $\mathfrak{u}$, $\mathfrak{v}$, $\mathfrak{w}$ stand for the following terms:

$$\mathfrak{u} = (1+\vartheta_1)\bar{u} + \psi_1\bar{v} + \chi_1\bar{w},$$

$$\mathfrak{v} = \vartheta_2\bar{u} + (1+\psi_2)\bar{v} + \chi_2\bar{w},$$

$$\mathfrak{w} = \left(\frac{d\hat{u}}{dz} + x\frac{d\vartheta_1}{dz} + y\frac{d\vartheta_2}{dz}\right)\bar{u} + \left(\frac{d\hat{v}}{dz} + x\frac{d\psi_1}{dz} + y\frac{d\psi_2}{dz}\right)\bar{v} + \left(\frac{d\hat{w}}{dz} + x\frac{d\chi_1}{dz} + y\frac{d\chi_2}{dz}\right)\bar{w}. \qquad \text{(VI.72)}$$

As in the preceding section, one can introduce the three angles α, β, γ, in terms of which $\vartheta_1, \vartheta_2, \psi_1, \psi_2, \chi_1, \chi_2$ are given by (VI.60) and the derivatives $\frac{d\hat{u}}{dz}, \frac{d\hat{v}}{dz}, \frac{d\hat{w}}{dz}$ by (VI.61). Then k_{xx}, k_{yy}, k_{xy} are given by (VI.62) and $\nu_{xx}, \nu_{yy}, \nu_{xy}$ by (VI.63).

This enables us to neglect all terms which contain $\nu_{xx}, \nu_{yy}, \nu_{xy}$ in the last of equations (VI.71).

Similarly, one may neglect the terms which contain $\frac{d\hat{\varepsilon}_{xx}}{dz}, \frac{d\hat{\varepsilon}_{yy}}{dz}, \frac{d\hat{\varepsilon}_{xy}}{dz}$ in (VI.71).

We now want to show that, in addition to these approximations, the possibility of which was established in § 50, one can also neglect all the underlined terms in (VI.71).

Substituting into

$$X = \bar{u}\frac{d\vartheta_1}{dz} + \bar{v}\frac{d\psi_1}{dz} + \bar{w}\frac{d\chi_1}{dz},$$

$$Y = \bar{u}\frac{d\vartheta_2}{dz} + \bar{v}\frac{d\psi_2}{dz} + \bar{w}\frac{d\chi_2}{dz}, \qquad \text{(VI. 73)}$$

the expressions for $\vartheta_1, \vartheta_2, \psi_1, \psi_2, \chi_1, \chi_2$ in accordance with (VI.60) and replacing the derivatives $\frac{d\alpha}{dz}, \frac{d\beta}{dz}, \frac{d\beta}{dz}$ in accordance with (VI.62), we obtain

$$X = \left[\bar{u}\frac{d\hat{u}}{dz} + \bar{v}\frac{d\hat{v}}{dz} + \bar{w}\left(1 + \frac{d\hat{w}}{dz}\right)\right]k_{xx} - k_{xy}\mathfrak{v},$$

$$Y = \left[\bar{u}\frac{d\hat{u}}{dz} + \bar{v}\frac{d\hat{v}}{dz} + \bar{w}\left(1 + \frac{d\hat{w}}{dz}\right)\right]k_{yy} + k_{xy}\mathfrak{u}. \qquad \text{(VI. 74)}$$

Using (VI.72) and (VI.73), equations (VI.74) may be rewritten in the form

$$X = (\mathfrak{w} - xX - yY)\,k_{xx} - \mathfrak{v}\,k_{xy},$$

$$Y = (\mathfrak{w} - xX - yY)\,k_{yy} + \mathfrak{u}\,k_{xy}. \qquad \text{(VI. 75)}$$

Hence

$$X = \frac{\mathfrak{w}\,k_{xx} - \mathfrak{v}\,(1 + yk_{yy})\,k_{xy} - \mathfrak{u}\,yk_{xx}k_{xy}}{1 + xk_{xx} + yk_{yy}},$$

$$Y = \frac{\mathfrak{w}\,k_{yy} + \mathfrak{u}\,(1 + xk_{xx})\,k_{xy} + \mathfrak{v}\,xk_{yy}k_{xy}}{1 + xk_{xx} + yk_{yy}}. \qquad \text{(VI. 76)}$$

Since xk_{xx}, yk_{yy} are of the same order of magnitude as the strain components, we have

$$X \approx \mathfrak{w}\,k_{xx} - \mathfrak{v}\,k_{xy},$$

$$Y \approx \mathfrak{w}\,k_{yy} + \mathfrak{u}\,k_{xy}. \qquad \text{(VI. 77)}$$

Thus the underlined terms in the fourth and fifth of equations (VI.71) are of the same order of magnitude as the products of $\mathfrak{u}$, $\mathfrak{v}$, $\mathfrak{w}$ by the curvature parameters of the axis of the rod in the strained state. But the supplementary displacements $\mathfrak{u}$, $\mathfrak{v}$, $\mathfrak{w}$ are always very small compared to the lateral dimensions of rods. Hence, since the products of the curvature parameters k_{xx}, k_{yy}, k_{xy} of the axis of the rod by its lateral dimensions are of the same order of magnitude as the strain components, one may conclude that X and Y are always small in comparison with the elongations and shears. From this it follows that the underlined terms in the fourth and fifth of equations (VI.71) can be neglected. Similarly, one can show that the underlined terms in the last of equations (VI.71) can be omitted.

With these approximations, we arrive at the following expressions for the strain components of a thin initially prismatic bar:

$$\begin{aligned}
&\varepsilon_{xx} = \hat{\varepsilon}_{xx} + \frac{\partial \mathfrak{u}}{\partial x}, \quad \varepsilon_{yy} = \hat{\varepsilon}_{yy} + \frac{\partial \mathfrak{v}}{\partial y}, \\
&\varepsilon_{xy} = \hat{\varepsilon}_{xy} + \frac{\partial \mathfrak{u}}{\partial y} + \frac{\partial \mathfrak{v}}{\partial x}, \\
&\varepsilon_{xz} = \hat{\varepsilon}_{xz} + yk_{xy} + \frac{\partial \mathfrak{u}}{\partial z} + \frac{\partial \mathfrak{w}}{\partial x}, \\
&\varepsilon_{yz} = \hat{\varepsilon}_{yz} - xk_{xy} + \frac{\partial \mathfrak{v}}{\partial z} + \frac{\partial \mathfrak{w}}{\partial y}, \\
&\varepsilon_{zz} = \hat{\varepsilon}_{zz} + xk_{xx} + yk_{yy} + \frac{\partial \mathfrak{w}}{\partial z}.
\end{aligned} \tag{VI. 78}$$

Here $\mathfrak{u}$, $\mathfrak{v}$, $\mathfrak{w}$ are functions of all three coordinates x, y, z, with the following properties:

a) They are very small in comparison with the lateral dimensions of the rod and their derivatives are of the same order of magnitude as the strain components. Due to this fact, products of pairs of derivatives and the product of a derivative by a quantity of the order of magnitude of the strain components can be neglected.

b) For $x = 0$, $y = 0$,

$$\mathfrak{u} = \mathfrak{v} = \mathfrak{w} = \frac{\partial \mathfrak{u}}{\partial x} = \frac{\partial \mathfrak{u}}{\partial y} = \frac{\partial \mathfrak{v}}{\partial x} = \frac{\partial \mathfrak{v}}{\partial y} = \frac{\partial \mathfrak{w}}{\partial x} = \frac{\partial \mathfrak{w}}{\partial y} = 0,$$

which follows from (VI.49) and (VI.72).

Adjusting the supplementary displacements $\mathfrak{u}$, $\mathfrak{v}$, $\mathfrak{w}$, one can bring equations (VI.78) into agreement with boundary conditions on the lateral surface of the rod.

§ 52. Pure Torsion

Let us subject a rod to a uniform torsion along its whole length.

Then

$$\alpha=\beta=k_{xx}=k_{yy}=0;\ \gamma=\tau z, \tag{VI.79}$$

where

$$\tau=k_{xy} \tag{VI.80}$$

is a constant.

We shall assume furthermore that

$$\hat{\varepsilon}_{xx}=\hat{\varepsilon}_{yy}=\hat{\varepsilon}_{xy}=\hat{\varepsilon}_{xz}=\hat{\varepsilon}_{yz}=\hat{\varepsilon}_{zz}=0 \tag{VI.81}$$

holds, i. e., we neglect the strains which are uniformly distributed along the cross-section of the rod.

Furthermore, from (VI.79) and (VI.81) we obtain

$$\hat{u}=\hat{v}=\hat{w}=0. \tag{VI.82}$$

Taking all this into account, we obtain

$$\begin{aligned} u &= -x(1-\cos\tau z)-y\sin\tau z+\bar{u}, \\ v &= +x\sin\tau z-y(1-\cos\tau z)+\bar{v}, \\ w &= \bar{w}\ , \end{aligned} \tag{VI.83}$$

and

$$\begin{aligned} &\varepsilon_{xx}=\frac{\partial\mathfrak{u}}{\partial x},\quad \varepsilon_{yy}=\frac{\partial\mathfrak{v}}{\partial y},\quad \varepsilon_{xy}=\frac{\partial\mathfrak{u}}{\partial y}+\frac{d\mathfrak{v}}{\partial x}, \\ &\varepsilon_{xz}=y\tau+\frac{\partial\mathfrak{u}}{\partial z}+\frac{\partial\mathfrak{w}}{\partial x}, \\ &\varepsilon_{yz}=-x\tau+\frac{\partial\mathfrak{v}}{\partial z}+\frac{\partial\mathfrak{w}}{\partial y}, \\ &\varepsilon_{zz}=\frac{\partial\mathfrak{w}}{\partial z}. \end{aligned} \tag{VI.84}$$

If we now set

$$\mathfrak{u}=\mathfrak{v}=0,\ \mathfrak{w}=\tau\varphi(x,\ y), \tag{VI.85}$$

then the expressions (VI.84) become

$$\varepsilon_{xx} = \varepsilon_{yy} = \varepsilon_{xy} = \varepsilon_{zz} = 0,$$

$$\varepsilon_{xz} = \tau\left(\frac{\partial\varphi}{\partial x} + y\right), \quad \varepsilon_{yz} = \tau\left(\frac{\partial\varphi}{\partial y} - x\right), \qquad \text{(VI.86)}$$

and thus yield the equations of Saint-Venant's theory of torsion. Hence Saint-Venant's theory is a special case of the general equations (VI.78).

We note that the displacements of points of the twisted rod are determined in our discussion by the expressions (VI.83) and not by the expressions of the classical theory:

$$u = -yz\tau, \quad v = xz\tau, \quad w = \tau\varphi(x, y). \qquad \text{(VI.87)}$$

These equations may be obtained from (VI.83) by supposing that $\tau z \ll 1$, i. e., by assuming that the angles of rotation under torsion are negligibly small compared to unity. It is clear from the above that this restriction is not necessary for Saint-Venant's theory.

§ 53. The Final Expressions for the Strain Components of a Thin Rod

It can be foreseen that in the general case of the deformation of a rod (when it is subjected not only to twisting but also to bending), equations (VI.66) will be inadequate. Indeed, as was shown at the end of § 49, they were already inadequate in the case of pure torsion. It may furthermore be foreseen that the necessary corrections which must be introduced into these equations will have the same character in the general case as they do in the case of pure torsion. More specifically, these corrections must allow us to remove the stresses which twist the rod and act on its surface, which arise unavoidably in using equations (VI.66) (for rods of noncircular cross-sections).

Hence one should attempt to construct a general theory of deformation of thin rods by setting, as in the preceding section,

$$\mathfrak{u} = \mathfrak{v} = 0, \quad \mathfrak{w} = k_{xy}(z) \cdot \varphi(x,y). \tag{VI.88}$$

Equations (VI.78) then assume the form

$$\begin{gathered}
\varepsilon_{xx} = \hat{\varepsilon}_{xx}, \quad \varepsilon_{yy} = \hat{\varepsilon}_{yy}, \quad \varepsilon_{xy} = \hat{\varepsilon}_{xy}, \\
\varepsilon_{xz} = \hat{\varepsilon}_{xz} + \left(\frac{\partial \varphi}{\partial x} + y\right) k_{xy}, \quad \varepsilon_{yz} = \hat{\varepsilon}_{yz} + \left(\frac{\partial \varphi}{\partial y} - x\right) k_{xy}, \\
\varepsilon_{zz} = \hat{\varepsilon}_{zz} + x k_{xx} + y k_{yy} + \varphi \cdot \frac{dk_{xy}}{dz}.
\end{gathered} \tag{VI.89}$$

These expressions are actually adequate for the problem at hand. With them as a basis, a consistent theory of deformation of flexible rods can be constructed, restricted only by the assumption that the elongations and shears are negligible compared to unity. The error in this theory can be estimated by comparing the elongations and shears with the angles of rotation since, in deriving (VI.78), we systematically neglected the former in comparison with the latter.

§ 54. Conclusion

Ordinarily, the theory of deformation of flexible bodies (plates, shells, rods) is developed by making certain assumptions which immediately reduce the problem to a two-dimensional one (in the case of plates and shells) or a one-dimensional problem (in the case of rods). However, with such assumptions one necessarily loses sight of the connection between the theory of plates, shells, and rods and the general theory of elasticity. In view of this, many people consider the theory of flexible bodies as a kind of hypothetical superstructure over the general theory of elasticity, as a foreign element in it.

Only in this manner can one probably explain why most contemporary books on the theory of elasticity omit all mention of the problem of deformation of flexible bodies, which is of such practical importance. An attempt was made in Love's book to relate the "hypotheses" of the theory of flexible bodies to the general theory of deformation.

But special work in this direction was carried out by B. G. Galerkin, in whose papers the classical theory of shells and plates truly became a branch of the general theory of elasticity.

The basic idea championed by B. G. Galerkin was that the problems of the bending of plates and shells must always be examined in the context of the general theory. This simple but profound idea was responsible to a large extent for the successful development of the theory of plates and shells in the Soviet Union and turned out to be fruitful not only in the case of thick plates and shells, but also in the case of thin plates and shells.

It is natural to extend this idea to the nonlinear theory of elasticity, since one can expect that many results of this theory may be systematized by starting out from the general equations. The present chapter was an attempt to give a uniform method for investigating the deformation of flexible bodies on the basis of the general nonlinear theory of deformations. It was our aim to clarify, with the aid of the general equations, those "hypotheses'' on which the theory of plates, shells, and rods is ordinarily based, and to examine, from a uniform point of view, all these problems, which are ordinarily treated separately in spite of their common features.

BIBLIOGRAPHY

Almansi, E.

1. "L'ordinaria teoria dell'Elasticità e la teoria delle deformazioni finite." Rendiconti della R. Accademia dei Lincei (5), vol. 26 (1917) (2nd sem.), pp. 3-8.

Ariano, R.

1. "L'isotropia nelle deformazioni finite. I. L'isotropia in senso largo." Rendiconti dell'Istituto Lombardo (2), vol. 66 (1933), pp. 92-104.
2. "L'isotropia nelle deformazioni finite. II. Isotropia in senso ristretto." Rendiconti dell' Istituto Lombardo (2), vol. 66 (1933), pp. 207-220.
3. "Sulla resistenza a torsione della gomma elastica e dell' ebonite." Il Politecnico, Milan, vol. 81 (1933), pp. 271-284.
4. "Le deformazioni finite." R. Scuola d' Ingegneria, Atti, Ricerche, Studi 18, Milan, Hoepli, 1935.

Bergeot, P.

1. "Sur la forme de l'énergie de déformation dans les déformations finies." Comptes Rendus de l'Académie des Sciences, Paris, vol. 218 (1944), pp. 824-825.

Biezeno, C. B.

1. "Über die Bestimmung der Durchschlagskraft einer schwachgekrümmten kreisförmigen Platte." Zeitschrift für angewandte

Mathematik und Mechanik, vol. 15 (1935), pp. 10-22.

Biezeno, C. B., and Hencky, H.

1. "On a general theory of elastic stability." I, II. Koninklijke Akademie van Wetenschappen te Amsterdam, Proceedings, vol. 31 (1928), pp. 569-592.
2. "On the general theory of elastic stability." Koninklijke Akademie van Wetenschappen te Amsterdam, Proceedings, vol. 32 (1929), pp. 444-456.

Biot, M. A.

1. "Increase of torsional stiffness of a prismatical bar due to axial tension." Journal of Applied Physics, vol. 10 (1939), pp. 860-864.
2. "Theory of elasticity with large displacements and rotations." Proceedings of the Fifth International Congress of Applied Mechanics, 1938, Cambridge, Mass., 1939, pp. 117-122.
3. "Elastizitätstheorie zweiter Ordnung mit Anwendungen." Zeitschrift für angewandte Mathematik und Mechanik, vol. 20 (1940), pp. 89-99.

Birch, F.

1. "The effect of pressure upon the elastic parameters of isotropic solids, according to Murnaghan's theory of finite strain." Journal of Applied Physics, vol. 9 (1938), pp. 279-288.

Blinnik, S., Malinin, N., Ponomarev, S., Feodosiev, V.

1. "New methods for the calculation of springs." Mashgiz, 1946. (Russian.)

Blokh, V.

1. "On finite deformations." Nauchn. zap. Khark. mekh. mach. inst., vol. 5 (1940), pp. 3-26. (Russian.)

Bonvicini, D.

1. "Sulle deformazioni non infinitesime." Rendiconti della R. Accademia dei Lincei (6), vol. 16 (1932), pp. 607-612.

Brillouin, M.

1. "Déformations homogènes finies. Energie d'un corps isotrope." Comptes Rendus de l'Académie des Sciences, Paris, vol. 112 (1891), pp. 1500-1502.

Bromberg, E., and Stoker, J. J.

1. "Nonlinear theory of curved elastic sheets." Quarterly of Applied Mathematics, vol. 3 (1945), pp. 246-265.

Bubnov. I. G.

1. "Structural mechanics of a ship," vol. I, 1913; vol. II, 1914, St. Petersburg. (Russian.)

Bulffinger, G. B.

1. "De solidorum resistentia specimen." Commentarii Academiae Scientiarum Imperialis Petropolitanae, vol. 4 (1729), pp. 164-181.

Burgatti, P.
1. "Sulle deformazioni finite dei corpi continui." Memorie della R. Accademia della Scienze dell' Istituto di Bologna. Classe di scienze fisiche, ser. 7, vol. 1 (1914), pp. 237-244.

Ceruti, G.
1. "Sopra un'estensione della teoria elastica alla seconda approssimazione." Rendiconti dell' Istituto Lombardo (2), vol. 65 (1932), pp. 997-1012.

Chien, W.
1. "The intrinsic theory of thin shells and plates." I. "General theory." Quarterly of Applied Mathematics, vol. 1 (1944), pp. 297-327.
2. "The intrinsic theory of thin shells and plates." II. "Application to thin plates." Quarterly of Applied Mathematics, vol. 2 (1944), pp. 43-59.
3. "The intrinsic theory of thin shells and plates." III. "Application to thin shells." Quarterly of Applied Mathematics, vol. 2 (1944), pp. 120-135.

Cosserat, E., and Cosserat, F.
1. "Sur la théorie de l'élasticité. Premier Mémoire." Annales de la Faculté des Sciences de Toulouse, vol. 10 (1896), pp. 1-116.

Crudeli, U.
1. "Sopra le deformazioni finite. Le equazioni del De Saint-Venant." Rendiconti della R.

Accademia dei Lincei, vol. 20 (1911) (2nd sem.), pp. 306-308.

Dinnik, A.
1. "Stability of elastic systems." ONTI, 1935 (Russian.)

Duhem, P.
1. "Recherches sur l'élasticité. Première partie. De l'équilibre et du mouvement des milieux vitreux." Annales Scientifiques de l'École Normale Supérieure, (3), vol. 21 (1904), pp. 99-139.
2. "Recherches sur l'élasticité. Deuxième partie. Les milieux vitreux peu déformés." Annales Scientifiques de l'École Normale Supérieure, (3), vol. 21 (1904), pp. 375-414.
3. "Recherches sur l'élasticité. Troisième partie." Annales Scientifiques de l'École Normale Supérieure, (3), vol. 22 (1905), pp. 143-217.
4. "Recherches sur l'élasticité. Quatrième partie. Propriétés générales des ondes dans les milieux visqueux et non visqueux." Annales Scientifiques de l'École Normale Supérieure, (3), vol. 23 (1906), pp. 169-223.

Feodosiev, V. I.
1. "Large displacements and stability of a circular membrane with fine corrugations." Journal of Applied Mathematics and Mechanics (Akad. Nauk SSSR. Prikl. Mat. Mekh.) (N.S.), vol. 9 (1945), pp. 389-412 (Russian. English summary.)
2. "Calculation of thin clicking membranes." Journal of Applied Mathematics and Me-

Feodosiev, V. I. (cont.)
chanics (Akad. Nauk SSSR. Prikl. Mat. Mekh.) (N.S.), vol. 10 (1946), pp. 295-300. (Russian. English summary.)

Galerkin, B. G.
1. "Series solutions of some problems of elastic equilibrium of rods and plates." Vestnik Inzhenerov, vol. 1 (1915), pp. 879-908. (Russian.)

Gorgidze, A., and Rukhadze, A.
1. "Bending of a twisted bar by a couple." Bulletin of the Academy of Sciences of the Georgian SSR (Soobshchenia Akademii Nauk Gruzinskoi SSR), vol. 5 (1944), pp. 253-262 (Georgian. Russian summary.)

James, H. M., and Guth, E.
1. "Wave equations for finite elastic strain." Journal of Applied Physics, vol. 16 (1945), pp. 643-644.

Kachanov, L. M.
1. "On the mechanics of plastic solids." Journal of Applied Mathematics and Mechanics (Akad. Nauk SSSR. Prikl. Mat. Mekh.) (2), vol. 4, Nr. 3 (1940), pp. 37-42. (Russian. English summary.)
2. "Elastico-plastic condition of hard bodies." Journal of Applied Mathematics and Mechanics (Akad. Nauk SSSR. Prikl. Mat. Mekh.) (2), vol. 5 (1941), pp. 431-438. (Russian. English summary.)

Kappus, R.

1. "Zur Elastizitätstheorie endlicher Verschiebungen."

   I. Zeitschrift für angewandte Mathematik und Mechanik, vol. 19 (1939), pp. 271-285.

   II. Zeitschrift für angewandte Mathematik und Mechanik, vol. 19 (1939), pp. 344-361.

Karmán, Th. von, and Tsien, H.

1. "The buckling of spherical shells by external pressure." Journal of the Aeronautical Sciences, vol. 7 (1939), pp. 43-50.

Kirchhoff, G.

1. "Über die Gleichungen des Gleichgewichts eines elastischen Körpers bei nicht unendlich kleinen Verschiebungen seiner Theile." Sitzungsberichte der Kaiserlichen Akademie der Wissenschaften, Wien, Mathematisch-Naturwissenschaftliche Classe, vol. 9 (1852), pp. 762-773.
2. "Ueber das Gleichgewicht und die Bewegung eines unendlich dünnen elastischen Stabes," Journal für die reine und angewandte Mathematik, vol. 56 (1859), pp. 285-313.

Krylov, V. V.

1. "Plane problem of the theory of elasticity for finite displacements." Journal of Applied Mathematics and Mechanics (Akad. Nauk SSSR. Prikl. Mat. Mekh.), vol. 10 (1946), pp. 647-656. (Russian. English summary.)

Kutilin, D.

1. "Theory of finite deformations." Gostekhizdat.,1947, pp. 1-275. (Russian.)

Landau, L., and Lifshitz, E.

1. "Mechanics of continuous media," 1944. (Russian.)

Latshaw, E.

1. "An elastic theory for rubber." Journal of the Franklin Institute, vol. 234 (1942), pp. 63-73.

Leibenson, L.

1. "A course in the theory of elasticity." Gostekhizdat., 1947. (Russian.)

Locatelli, P.

1. "Sopra il teorema del minimo lavoro per corpi non perfettamente elastici." Atti Accad. Italia. Rend. Cl. Sci. Fis. Mat. Nat. (7), vol. 1 (1939), pp. 10-18.

Love, A. E. H.

1. "A treatise on the mathematical theory of elasticity." Fourth Edition, New York, Dover Publications, 1944.

Lur'ye, A.

1. "Generalization of a theorem of Castigliano" Trudy, Leningrad Polytech. Inst., No. 1, 1946. (Russian.)

Marcolongo, R.

1. "Le formule del Saint-Venant per le deformazioni finite." Rendiconti del Circolo

Marcolongo, R. (cont.)
Matematico di Palermo, vol. 19 (1905), pp. 151-155.

Marguerre, K.
1. "Über die Anwendung der energetischen Methode auf Stabilitätsprobleme." Jahrbuch der Deutschen Luftfahrtforschung, Flugwerk, 1938, pp. 433-443.
2. "Zur Theorie der gekrümmten Platte grosser Formänderung." Proceedings of the Fifth International Congress of Applied Mechanics, 1938, Cambridge, Mass., 1939, pp. 93-101.

Murnaghan, F. D.
1. "Finite deformations of an elastic solid." American Journal of Mathematics, vol. 59 (1937), pp. 235-260.
2. "On the theory of the tension of an elastic cylinder." Proceedings of the National Academy of Sciences, U. S. A., vol. 30 (1944), pp. 382-384.

Mushtari, Kh.
1. "A certain generalization of the theory of thin shells." Izv. Fiz. Mat. Obshchestva pri Kazanskom Universitete, vol. 11, ser. 8 (1938). (Russian.)

Newing, S. T., and Shepherd, W. M.
1. "Finite strain: dislocation solutions." Philosophical Magazine, vol. 26 (1938), pp. 557-569.

Novozhilov, V.
1. "General theory of stability of thin shells."

Novozhilov, V. (cont.)

Comptes Rendus (Doklady) de l'Académie des Sciences de l'URSS (N.S.), vol. 32 (1941), pp. 316-319. (Russian.)

Panov, D. J.

1. "Application of Galerkin's method to some nonlinear problems of the theory of elasticity." Journal of Applied Mathematics and Mechanics (Akad. Nauk SSSR. Prikl. Mat. Mekh.), vol. 3 (1939), pp. 139-142. (Russian. English summary.)
2. "Secondary effects of the twisting of an elliptical cylinder." Trudy, Central Aero-Hydrodynamical Institute, 459 (1939). (Russian.)
3. "On large deflections of slightly corrugated circular membranes." Journal of Applied Mathematics and Mechanics (Akad. Nauk SSSR. Prikl. Mat. Mekh.), vol. 5 (1941), pp. 303-318. (Russian. English Summary.)

Papkovich, P. F.

1. "Structural mechanics of a ship," vol. 2, Oborongiz, Leningrad, 1941, pp. 1-159. (Russian.)

Popov, E.

1. "The theory and calculation of flexible, elastic parts." Izd. LKVVIA, 1947. (Russian.)

Riz, P. M.

1. "On some phenomena connected with the twisting of a circular cylinder." Trudy, Central Aero-Hydrodynamical Institute 408 (1939). (Russian.)

Riz, P. M. (cont.)

2. "Elastic constants in the nonlinear theory of elasticity." Journal of Applied Mathematics and Mechanics (Akad. Nauk SSSR. Prikl. Mat. Mekh.), vol. 11 (1947), pp. 493-494. (Russian. English summary.)
3. "The theory of elasticity for large deformations exceeding the limit of proportionality." Comptes Rendus (Doklady) de L'Académie des Sciences de l'URSS(N.S.), vol. 59 (1948), pp. 223-225. (Russian.)

Rukhadze, A. K.

1. "The bending of an extended prismatic rod." Bulletin of the Academy of Sciences of the Georgian SSR (Soobshchenia Akademii Nauk Gruzinskoi SSR), vol. 2 (1941), pp. 609-617. (Georgian. Russian summary.)
2. "Influence of transverse force on torque in the bending of a bar." Journal of Applied Mathematics and Mechanics (Akad. Nauk SSSR. Prikl. Mat. Mekh.), vol. 11 (1947), pp. 351-356. (Russian. English summary.)

Sadowsky, M.

1. "Nonlinear springs." Journal of the Franklin Institute, vol. 240 (1945), pp. 469-476.

Saint-Venant, B. de

1. "Mémoire sur l'équilibre des corps solides, dans les limites de leur élasticité, et sur les conditions de leur résistance, quand les déplacements éprouvés par leurs points ne sont pas très-petits." Comptes Rendus de l'Académie des Sciences, Paris, vol. 24 (1847), pp. 260-263.

Seth, B. R.
1. "Finite strain in elastic problems." Philosophical Transactions of the Royal Society (London), ser. A, vol. 234 (1935), pp. 231-264.
2. "Finite strain in aelotropic elastic bodies." I. Bulletin of the Calcutta Mathematical Society, vol. 37 (1945), pp. 62-68.

Shepherd, W. M., and Seth, B. R.
1. "Finite strain in elastic problems." Proceedings of the Royal Society, London, vol. 156 (1936), pp. 171-192.

Shtaerman, I., and Pikovski, A.
1. "Methods of calculating constructions for stability," Ukrgizmestprom, 1939. (Russian.)

Signorini, A.
1. "Sulle deformazioni finite dei sistemi a trasformazioni reversibili." Rendiconti della R. Accademia dei Lincei (6), vol. 18 (1933), pp. 388-395.

Sokolnikoff, I. S.
1. "Mathematical theory of elasticity." New York, McGraw-Hill, 1946.

Sokolov, P.
1. "On stresses in compressed plates after loss of stability." Trudy NISS, issue 7, 1932. (Russian.)

Sternberg, E.
1. "Nonlinear theory of elasticity with small deformations." Journal of Applied Me-

Sternberg, E. (cont.)
chanics, vol. 13 (1946), pp. A53-A60.

Timoshenko, S. I.
1. "On the stability of elastic systems." Izv. Kiev. Polytech. Inst., 1910. (Russian.)

Trefftz, E.
1. "Handbuch der Physik", vol. 6, "Mechanik der elastischen Körper," Berlin, Springer, 1928.
2. "Zur Theorie der Stabilität des elastischen Gleichgewichts." Zeitschrift für angewandte Mathematik und Mechanik, vol. 13 (1933), pp. 160-165.

Voigt, W.
1. "Ueber eine anscheinend notwendige Erweiterung der Theorie der Elasticität." Nachrichten von der Königlichen Gesellschaft der Wissenschaften zu Göttingen 1893, pp. 534-552; 1894, pp. 33-42.

Yasinski, F. S.
1. "On resistance to lateral buckling." Scientific papers of F. S. Yasinski, vol. 1, St. Petersburg, 1902, p. 145. (Russian.)

Zvolinski, N. V.
1. "Torsion of a bar extended by constant mass forces." Journal of Applied Mathematics and Mechanics (Akad. Nauk SSSR. Prikl. Mat. Mekh.) (N.S.), vol. 3 (1939), pp. 3-32. (Russian. English summary.)

Zvolinski, N. V., and Riz, P. M.

1. "Hooke's law for finite displacements." Izvestiya Akademii Nauk SSSR, Otdeleniye tekhnicheskikh nauk, 1938, pp. 17-20. (Russian.)
2. "On some problems in the nonlinear theory of elasticity." Journal of Applied Mathematics and Mechanics (Akad. Nauk SSSR. Prikl. Mat. Mekh.) (N.S.), vol. 2 (1939), pp. 417-426. (Russian. English summary.)
3. "General solution of torsion problems in the nonlinear theory of elasticity." Journal of Applied Mathematics and Mechanics (Akad. Nauk SSSR. Prikl. Mat. Mekh.) (N.S.), vol. 7 (1943), pp. 149-154. (Russian.English summary.)

INDEX